Seaweeds of India

The Diversity and Distribution of Seaweeds of the Gujarat Coast

Developments in Applied Phycology 3

Series Editor:
Michael A. Borowitzka
School of Biological Sciences & Biotechnology
Murdoch University, Murdoch, Western Australia

For other titles published in this series, go to
www.springer.com/series/7591

Bhavanath Jha • C. R. K. Reddy
Mukund C. Thakur • M. Umamaheswara Rao

Seaweeds of India

The Diversity and Distribution of Seaweeds of the Gujarat Coast

CSMCRI

GSBTM

Springer

Bhavanath Jha
Central Salt and Marine Chemicals
Research Institute
Council of Scientific
and Industrial Research (CSIR)
Gijubhai Badheka Marg
Bhavnagar 364 002, Gujarat
India

C.R.K. Reddy
Central Salt and Marine Chemicals
Research Institute
Council of Scientific
and Industrial Research (CSIR)
Gijubhai Badheka Marg
Bhavnagar 364 002, Gujarat
India

Mukund C. Thakur
Central Salt and Marine Chemicals
Research Institute
Council of Scientific
and Industrial Research (CSIR)
Gijubhai Badheka Marg
Bhavnagar 364 002, Gujarat
India

M. Umamaheswara Rao
Andhra University
Visakhapatnam 530003
India

ISBN 978-90-481-2487-9 e-ISBN 978-90-481-2488-6
DOI 10.1007/978-90-481-2488-6
Springer Dordrecht Heidelberg London New York

Library of Congress Control Number: 2009927715

Printed on acid-free paper

Springer is part of Springer Science+Business Media (www.springer.com)

Foreword

India has 7,500 km of coastline with diverse habitats and rich biota. Coastal ecosystems, unfortunately, are experiencing wide range of pressures due to siltation, eutrophication, coastal development, aquaculture and climate change. Those species that adapt to these pressures will expand their living boundaries while others may fade away. Accordingly, the study of coastal biodiversity is of great concern globally and constitutes an important element of global change research.

Gujarat has 1,600 km of coastline, reportedly with rich diversity of seaweeds. Previously published accounts on seaweed biodiversity were mainly in the form of checklists, the earliest among these being the checklist of Krishnamurthy and Joshi prepared in the early 1970s. The more recent checklists are based almost entirely on secondary information.

The present book entitled *Seaweeds of India – The Diversity and Distribution of Seaweeds of Gujarat Coast* is a timely publication based wholly on primary data. Data were collected through extensive and systematic field studies conducted by the authors during different seasons over a three year period. The authors collected nearly 200 species of seaweeds belonging to 100 genera of Chlorophyta, Phaeophyta and Rhodophyta. Twenty-four of the species are new to Gujarat coast and three are new to Indian waters. The book contains high-quality images of the different species in their existing habitats. Brief taxonomical descriptions, together with information on ecology, distribution, seasonality and abundance, are covered for each of the species. Where applicable, the economic importance of the seaweed is also highlighted along with references. Molecular taxonomy was utilised to establish the correct nomenclature in certain ambiguous cases and also for those seaweeds which hold out promise as source of high value products such as agarose.

The book gains critical importance from the fact that the state of Gujarat is witnessing rapid development – which will no doubt have a bearing on the coastal environment – and the baseline data on seaweed resources would be useful to understand changes that may arise from physical changes in the environment as also pollution load and climate change. It is pertinent to note that, although Gujarat was largely unaffected by the Asian Tsunami of 2004, the study reflects the post-Tsunami scenario. Finally, with the recent introduction of mariculture in Indian waters and the expectation that commercial cultivation will be undertaken in Gujarat coast as well, the present work will enable accurate assessment of the impact of such cultivation.

I applaud the authors for their commendable work and am confident that the contents of this book will appeal to a large cross section of people engaged in coastal research and development.

Bhavnagar, Gujarat, India

Pushpito K. Ghosh
Director
Central Salt and Marine Chemicals Research Institute

Preface

India has the highest record of seaweed species from the Indian Ocean region. The country has a coastline of 7,500 km with diverse habitats supporting rich seaweed biodiversity.

The marine algal flora of Indian coast is largely known from the works of M.O.P. Iyengar and F. Børgesen who carried out extensive studies on seaweed collections from the Indian coast. Subsequently, F. Thivy, K. Srinivasan, V. Krishnamurthy, T.V. Desikachary, Sreenivasa Rao, Umamaheswara Rao and many others built on this knowledge and published detailed floristic accounts of Indian benthic marine algae. K. Srinivasan was the first phycologist to publish a book entitled *Icons of Indian Marine Algae* in 1969 enlisting 50 species collected by him from different Indian shores. This book has, unfortunately, been out of print for many years. The subsequent efforts in this direction focused mainly on preparation of checklists of Indian marine algae with geographical distribution and bibliographic references. The first such checklist was published by Krishnamurthy and Joshi in 1970 and has been widely referred to. Subsequently, several updated checklists have been prepared but these are largely based on secondary data and do not necessarily reflect the current scenario.

For the first time an effort has been made to bring out an illustrated book on the distribution and diversity of seaweeds of the Gujarat Coast based on systematic survey and collections made during December 2005–December 2008. This may be considered as the first volume under the umbrella of "Seaweeds of India", hopefully with more such volumes to follow. To make the identification process easier, we have provided high resolution colour photographs of species in their original habitat along with geographical coordinates of each location together with the ecology of the surrounding area. Key taxonomic characteristics of the species and their known uses are also included.

The book is intended to serve as a reliable and up-to-date reference source for students, teachers and researchers engaged in the area of marine algal biology. It will no doubt also serve as a valuable source of useful information for policy planners, environmentalists and marine-based industries. Indeed, the book provides a new baseline for industrially-strong Gujarat against which future changes in seaweed pattern arising from diverse causes such as pollution load, climate change and natural calamities can be monitored.

April 2009

Bhavanath Jha
C.R.K. Reddy
Mukund C. Thakur
M. Umamaheswara Rao

Acknowledgements

First of all we would like to express our sincere gratitude to Dr. Pushpito K. Ghosh, Director, Central Salt and Marine Chemicals Research Institute (CSIR) for mooting the concept behind the book and providing constructive ideas on the organization of data. The support and encouragement received throughout the project has also been most inspiriting.

We would also like to specially thank Dr. B K Ramavat, our former colleague, for guiding us to some field sites during the early phase of this work and for sharing GPS data on some of the collection spots. We also thank Mr. Vaibhav A. Mantri for sharing valuable information on occurrence and taxonomy of seaweeds. Special thanks are due to Mrs. Madhu Pareek and Dr. Avinash Mishra for molecular marker study of *Gracilaria* species.

The assistance rendered by Mr. Santlal Jaiswar, Junior Technical Assistant, during field visits and towards preparation and maintenance of the herbaria of the specimens collected in the course of the project is greatly appreciated. Many other scientists and research students in the Discipline of Marine Biotechnology and Ecology have extended support in one form or the other and we would like to thank them collectively.

We would like to thank our drivers for their endurance during long drives – sometimes even at odd hours – and for waiting patiently at each station till the job is done. Our sincere thanks are also due to all staff members in the departmental office and general administration for their co-operation and support which facilitated smooth execution of the work.

We are indebted to the Commanding Officer, Indian Coast Guard Station, Okha for providing us with a boat to visit islands in the Gulf of Kutch. We also thank The Port Officer, Gujarat Maritime Board, Okha and Executive Engineer, Public Works Department, Dwarka for extending their guest house facilities during the tenure of this study.

The publication of the book would not have been possible without the generous funding from our parent organization, Council of Scientific and Industrial Research (CSIR), New Delhi (NWP018) and the Gujarat State Biotechnology Mission (GSBTM), Government of Gujarat.

Last, but not the least, we would like to thank Ms. Suzanne Mekking, publishing editor, Aquatic Sciences, Springer Verlag for her keen interest in the publication of this book. Thanks are also due to anonymous reviewers for their valuable suggestions which helped in giving the book its final shape.

Contents

Introduction

Gujarat is situated on the north-western most part of peninsular India between 20° 1' to 24° 7' N and 68° 4' to 74° 4' E (Fig. 1). It has total area of 196,024 km^2 and a coastline of 1,650 km – the longest coastline of the country – with a continental shelf of 164,200 km^2 (35.3% of the country) and an Exclusive Economic Zone (EEZ) of 214,000 km^2 (9.9% of the country). Further, the state has two major gulfs namely Gulf of Khambat and Gulf of Kutch which embrace diverse coastal habitats as well as biota of ecological significance. The Gulf of Kutch is the biggest gulf on the west coast of India in the Arabian Sea and contains 42 islands fringing with corals and mangroves which provide congenial habitat for proliferation of seaweeds, nesting birds and animals. The coastal parts of both Gulf of Khambhat and Kutch region have extensive intertidal mud and sand flats, salt marshes, and mangrove vegetation, particularly in the river deltas of Gulf of Khambhat.

The Gujarat coast is composed of deccan traps and tertiary rocks and in places recent alluvium with lime stones with miliolite fossil forms, which are of Pleistocene age. At the Gulf of Khambhat, newer alluvium deposits are found. The tidal cycle on the Indian coast in general is semidiurnal, with two high and two low waters with unequal tidal amplitudes each day. The area of intertidal exposure depends on the tidal amplitudes and topography of the coast.

The long stretches of rocky, coralline and limestone substrata of both intertidal and shallow subtidal waters along Gujarat coast, particularly Saurashtra coast is dominated by enormous assemblage of diverse seaweed communities. The coasts where the intertidal zone is gently sloping towards the sea with vast stretches of rocky substrata often exposing and flooding during tidal cycles are found flourishing with wide variety of seaweeds either as open shore formations or sheltered formations (puddles, pools, lagoons, crevices, creeks, etc.). A pure sandy or muddy sea coast harbours few forms compared to hard substrata. The Gujarat coast perhaps surpasses all other coasts in India for the luxuriance and variety of seaweeds. The floristic variations in seaweed communities are controlled by several environmental factors including season, habitat, topography, duration of exposure, tidal amplitude and other biotic factors. The tidal amplitude of the Saurashtra coast, as compared to other parts of the west coast and the entire east coast of India, is very high. This results into maximum exposure of intertidal area, at places extending up to 1 km, especially during spring tides; as a result considerable growth of seaweeds in the intertidal and shallow subtidal regions is observed.

Although extensive marine algal collections have been made since the eighteenth century from India and Indian Ocean region, Iyengar (1927) was the first phycologist to publish the marine algal flora of the Indian coast. Subsequently, Børgesen (1935, 1937a, b) had a series of publications on Indian marine algae. Since then it has attracted the attention of many researchers and eventually resulted in publication of floristic accounts on Indian marine algae. Dixit (1968) recorded a total of 411 species of seaweeds, including those from Pakistan and Sri Lanka. Subsequently, Srinivasan (1969, 1973) published two volumes on the *Icones of Indian Marine Algae* describing 50 species, each based on collections made from the Indian coast. These volumes have, unfortunately, been out of print and at present are not available in the public domain. Later, Krishnamurthy and Joshi (1970) published "A Checklist of Indian Marine Algae" listing 520 species from the Indian coast including 153 species for Gujarat. This checklist was further updated by Untawale et al. (1983), Sahoo et al. (2001) and Oza and Zaidi (2001) and reported 620 species, 770 species and 844 species, respectively. However, these checklists are made by compiling the published

Fig. 1 Map of Gujarat showing collection locations selected for this study

Fig. 2 View of different seaweed collection sites along the Gujarat coast

records from various sources and do not ensure taxonomic authentication of species nor their present day availability.

Since there is no illustrated book on seaweeds of India, an effort was made to fill the void by bringing out this book through systematic collection, identification and documentation. The Saurashtra coast is rich in seaweed biodiversity, while South Gujarat and Kutch is largely muddy and only few common seaweed species were observed. The entire Gujarat coast was surveyed and fixed select localities for this study purpose (Fig. 2). Based on the relative abundance and seaweed diversity, stations, e.g. Gopnath, Navabander, Kotda, Sutrapada, Veraval, Adri, Porbandar, Dwarka, Vervala, Shivrajpur, Okha, Kandla, Mandvi, Koteswar and Samiyani Island (Gulf of Kutch), were selected for sampling of seaweeds. These stations were visited at monthly intervals from December 2005–December 2008 to collect as many species as possible to have a comprehensive mapping of seaweed diversity. For easy identification, colour photograph of each species in its natural habitat together with well spread fresh specimen and cross sections of thallus showing characteristic features are given. For each species, key taxonomic description, distribution with GPS data, abundance, seasonality, ecology and economic importance – wherever applicable – are also included. Molecular markers such as *cox*2 – 3 spacer, RUBISCO spacer, ITS1 and 18S rDNA were studied for all *Gracilaria* species described in the book and the respective NCBI GenBank accession numbers are provided.

Total 198 species belonging to 101 genera of seaweeds are described in this book. Among these, 24 species are new reports from Gujarat and 3 (*Solieria chordalis*, *Ahnfeltia plicata* and *Dictyopteris serrata*) from the Indian coast. The Rhodophyta contained the maximum number of species 109 belonging to 62 genera, followed by the Chlorophyta with 54 species belonging to 23 genera and the Phaeophyta with 35 species belonging to 16 genera. Economically important species such as *Gelidiella acerosa*, *G. micropterum*, *G. pusillum*, *Ahnfeltia plicata*, *Gracilaria dura*, *G. debilis*, *G. verrucosa*, *Hypnea musciformis*, *Meristotheca papulosa*, *Porphyra* sp, *Asparagopsis taxiformis*, *Sargassum tenerrimum*, *S. plagiophyllum*, *S. swartzii*, *Turbinaria ornata*, *Monostroma* sp. *Enteromorpha prolifera*, *E. compressa* and *E. tubulosa* are also present. *Caulerpa fastigiata* forma *delicatula*, *Enteromorpha ovata*, *Sargassum swartzii*, *Grateloupia indica*, *Helminthocladia clavadosii* forma *indica*, *Odontothalia veravalensis* and *Predaea feldmannii* var. *indica* are some of species endemic to the Gujarat coast.

Gujarat State is rich in marine bio-resources and is witnessing vibrant growth in industrial and infrastructural sectors. These developmental activities may influence coastal resources. In this context, this book will provide reliable baseline information for monitoring and evaluating the impact of rapid coastal development and to follow floristic changes with time.

We hope that the book will serve as an authentic field guide and reference material for students, teachers, researchers and policy planner interested in marine bio-resources and its sustainable utilization.

Chlorophyta

Ulotrichales, Monostromataceae

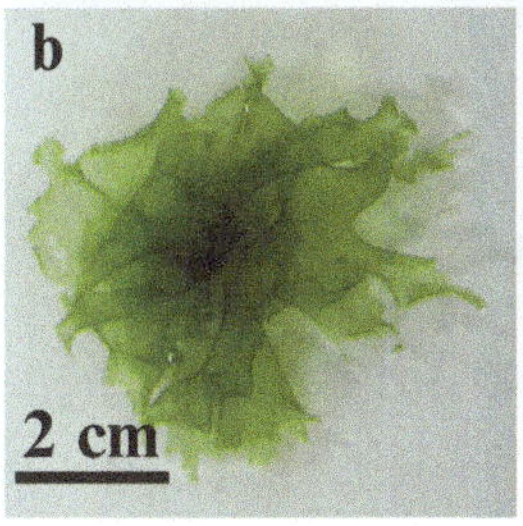

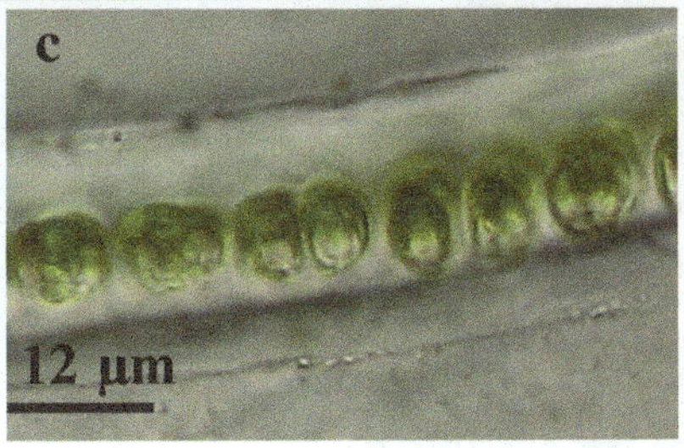

Distribution
Okha: N22° 28.55'; E 69° 04.33'
Sikka: N22° 26.23'; E 69° 49.70'
Samiani Island: N 22° 29.30'; E 69° 05.20'

Monostroma latissimum **Wittrock**

Thallus form and type	:	Small leafy fronds
Substratum	:	Rocks, gravels, pebbles, dead shells
Occurrence	:	Supra littoral and upper mid littoral zone
Season	:	January–May and August–November
Abundance	:	Scanty
Description	:	Plants usually grow as patchy tufts in upper mid littoral zone which submerges during high tide. Plants attached to rocks or barnacles as dense tufts (Fig. a), yellow green or bright green in colour, 2 – 3 cm tall, membranous with crisped margins (Fig. b); fronds monostromatic, 25 µm thick (Fig. c); cells usually oval, 12.5 µm broad, with laminate chloroplasts. The cells in surface view of the thallus are tetra or polygonal or oval in shape and embedded loosely in groups consisting of 2 – 4 cells each.
Use	:	*Monostroma* species are used for human consumption (as delicacy in soybean milk soups) in China, Japan and other countries. Annual production of *Monostroma* is about 1,250 t (dry wt) in Japan

B. Jha et al., *Developments in Applied Phycology*,
DOI 10.1007/978-90-481-2488-6_1, © Springer Science + Business Media B.V. 2009

Ulvales, Ulvaceae

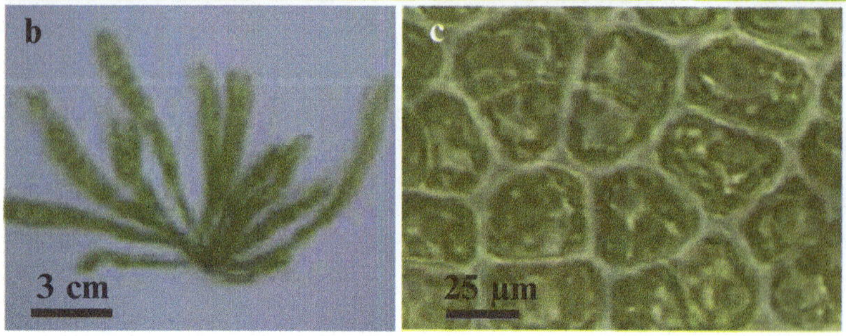

Distribution
Nava Bandar: N 22° 45.33'; E 71° 65.23'
Okha: N 22° 28.44'; E 69° 04.07'
Porbandar: N 21° 37.92'; E 69° 36.48'
Rajapara: N 20° 47.46'; E 71° 12.16'

***Enteromorpha compressa* (Linnaeus) Nees**

Thallus form and type	:	Tubulose and compressed
Substratum	:	Intertidal rocks, floating structures in the sea
Occurrence	:	Upper littoral and mid littoral zone of relatively calm waters
Season	:	Throughout the year
Abundance	:	Dense
Description	:	Plants yellowish green in colour, up to 11 cm long; fronds profusely branched from the stalk-like base, narrow, tubular at the base and gradually expanded and compressed above with obtuse or round apices (Figs. a, b); cells in surface view squarish or elongated and thin walled (Fig. c); plants growing at greater depths or in calm waters relatively longer. *E. compressa* is synonym of *Ulva compressa*.
Remark	:	*Enteromorpha* species are cultivated in Japan for edible application. The dry powder of *Enteromorpha* is mixed with fast food preparations and eaten in Japan and other countries

Ulvales, Ulvaceae

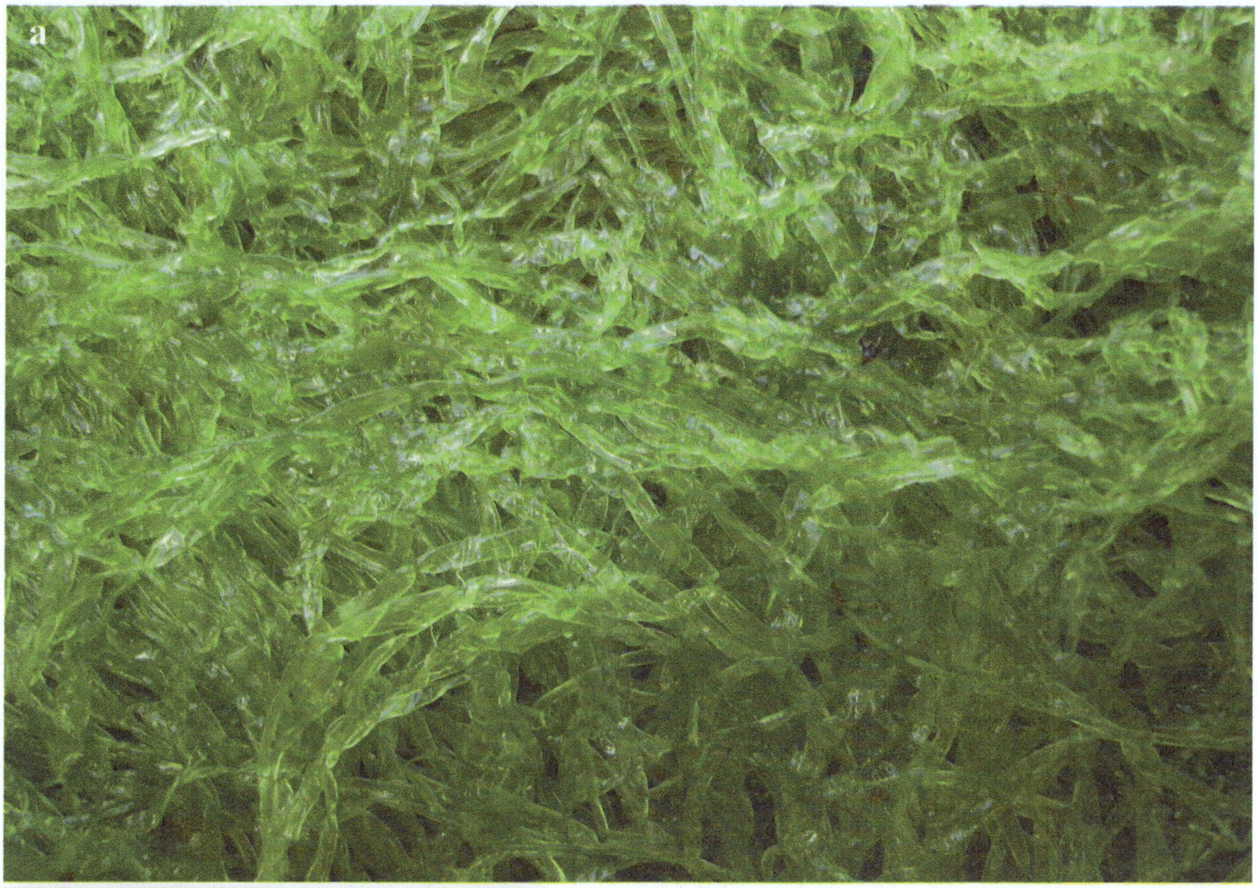

Distribution
Nava Bandar: N 22° 45.33'; E 71° 65.23'
Okha: N 22° 28.44'; E 69° 04.07'
Veraval: N 20° 54.87'; E 70° 20.83'

Enteromorpha flexuosa **(Wulf.) J. Agardh subsp.** ***flexuosa***

Thallus form and type	:	Tubular and flexuous
Substratum	:	Rocky, Pebbles or any floating structures in the sea
Occurrence	:	Upper mid littoral zone
Season	:	November–April
Abundance	:	Dense in selective habitats
Description	:	Plants yellowish green in colour, up to 30 cm long, attached by a small, round basal disk, fronds simple or branched, tubular with cylindrical stalks below and expanding blades above, becoming flexuous, ending in an obtuse apex (Figs. a, b), some times constricted at intervals; cells polygonal in surface view. *E. flexuosa* is a synonym of *Ulva flexuosa.*
Use	:	The dry powder is used in instant cup noodles in Japan. Recently its use as garnishing agent in snack foods is gaining world wide importance mainly due to availability of nourishing agents such as poly unsaturated fatty acids, minerals, vitamins, antioxidants and proteins in them

Ulvales, Ulvaceae

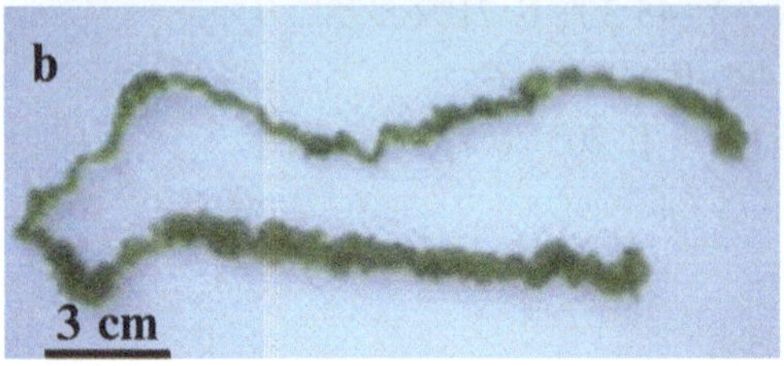

Distribution
Gopnath: N 21° 12.52'; E 72° 06.53'
Porrbandar: N 21° 38.24'; E 69° 35.81'

***Enteromorpha intestinalis* (Linnaeus) Nees**

Thallus form and type	:	Tubular and contorted
Substratum	:	Rocks, mudflats in mangrove environment and pebbles
Occurrence	:	Mid littoral zone in brackish water habitats (river estuaries)
Season	:	August–April
Abundance	:	Moderate in selective habitats especially in estuarine waters
Description	:	Plants simple or branched, light to yellow green in colour and up to 15 cm long (Fig. a), attached by basal rhizoidal portion and later become free-floating, mature specimens often inflated at intervals giving an intestine-like appearance, contorted and irregularly constricted (Fig. b). The name of this alga is currently regarded as synonym of *Ulva intestinalis*.
Use	:	The seaweed meal of this alga along with other green algae used as poultry feed in some countries. Annual production of this alga is about 1,400 t (dry wt) in Japan

Ulvales, Ulvaceae

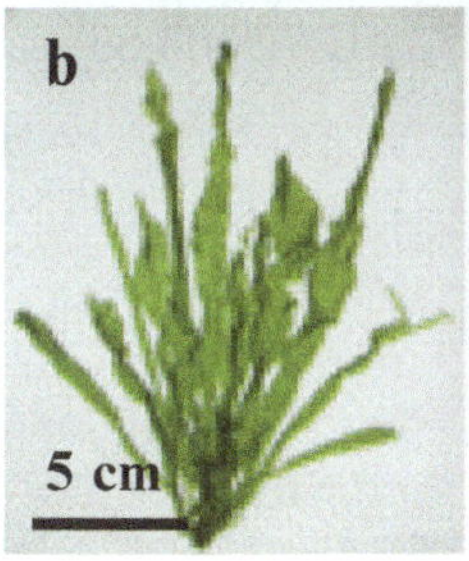

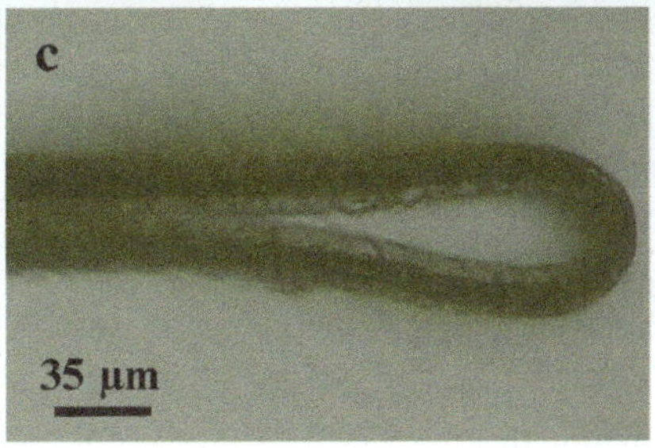

Distribution
Gopnath: N 21° 12.52'; E 72° 6.53'
Nava Bandar: N 22° 45.33'; E 71° 65.23'
Okha: N 22° 28.39'; E 69° 4.51'
Veraval: N 20° 54.87'; E 70° 20.83'

Enteromorpha linza **(Linnaeus) J. Agardh**

Thallus form and type	:	Tubular and leafy
Substratum	:	Rocks and on any floating structures in the sea
Occurrence	:	Upper littoral and mid littoral zone
Season	:	August–April
Abundance	:	Moderate in selective habitats in brackish environment
Description	:	Plants light to dark green in colour, up to 20 cm long, 3 cm wide (Figs. a, b), attached by basal discs, linear-lanceolate or oblanceolate; fronds flat and broad above, gradually tapering below, margins undulate; cells in surface view vertically elongated in basal portions and polygonal in the rest of the thallus; thallus distromatic and united at the middle part and separated only at the margins showing hollow one cell thick in nature (Fig. c). It is regarded as synonym of *Ulva linza.*
Use	:	The seaweed meal from green seaweeds is used as aquafeed and livestock feed in some countries

Ulvales, Ulvaceae

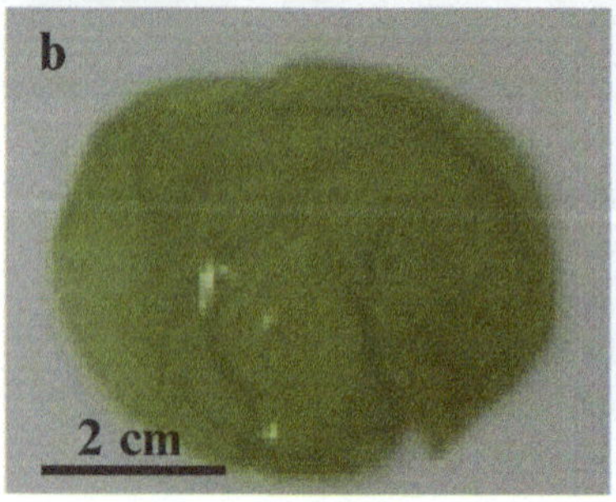

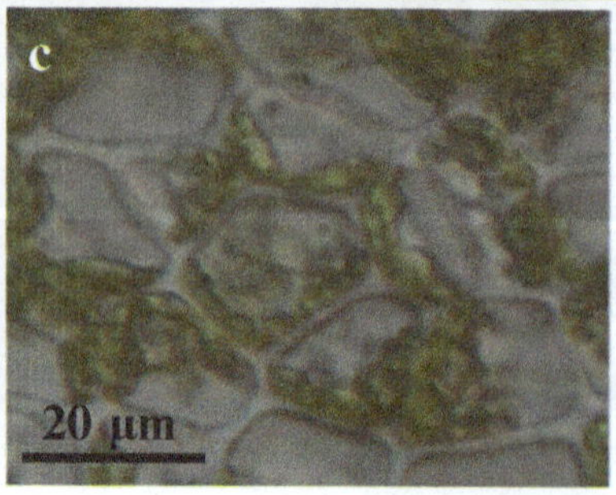

Distribution
Gopnath: N 21° 12.52'; E 72° 06.53'

Enteromorpha ovata **Thivy & Visalakshmi ex Joshi & Krishnamurthy**

Thallus form and type	:	Oval and unbranched
Substratum	:	Rocks, stones, shells, pebbles
Occurrence	:	Mid littoral zone, tide pools and placid waters
Season	:	November–April
Abundance	:	Dense in above mentioned location
Description	:	Plants yellowish green in colour, 0.5 – 4.0 cm long, 0.3 – 1.8 cm wide, attached by minute holdfast; fronds hollow, single, occasionally 4 – 5 from a common base, simple, flat, ovate (Figs. a, b); cells in surface view irregularly arranged through out the thallus, polygonal (Fig. c) with granular chloroplast. This alga is endemic to Indian coast and collected exclusively from above mentioned collection spot.

Ulvales, Ulvaceae

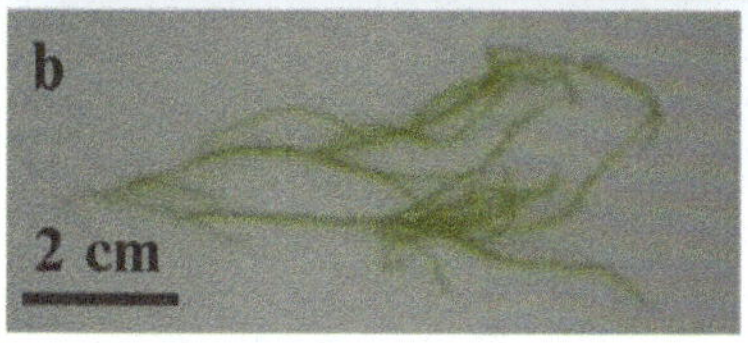

Distribution
Nava Bandar: N 22° 28.52'; E 69° 04.32'
Okha: N 22° 28.43'; E 69° 04.07'
Shivrajpur: N 22° 19.87'; E 68° 56.95'

***Enteromorpha prolifera* (Müller) J. Agardh**

Thallus form and type	:	Tubulose, branched and filamentous
Substratum	:	Rocks covered with sand or on any floating structures in the sea
Occurrence	:	Upper littoral and mid littoral zone of calm to moderately wave exposed shores
Season	:	August–April
Abundance	:	Dense in selective habitats
Description	:	Plants are green to dark green in colour, thallus with tubular branches that have numerous slender branchlets, up to 10 cm long, attached by means of a disk-like holdfast; fronds tubular, profusely branched with numerous slender branches (Figs. a, b); cells in surface view polygonal to sub-rectangular, always arranged in linear series in certain parts of the thallus. Recently it caused a green algal bloom in coastal waters of China.
Use	:	It is cultivated in Japan for edible application. Dry powder is used as garnishing agent in fast food preparations. Its use in snack foods is gaining importance in different countries

Ulvales, Ulvaceae

Distribution
Okha: N 22° 28.40'; E 69° 04.13'

Ulva conglobata **Kjellman**

Thallus form and type	:	Leafy and undivided
Substratum	:	Rocks or any other hard substrata, some times as epiphyte on other benthic seaweeds
Occurrence	:	Upper mid littoral zone, exposed and submerged rocks
Season	:	October–April
Abundance	:	Moderate
Description	:	Plants yellow to bright green in colour, entire, 2 – 4 cm tall and expanding as a globular mass (Figs. a, b); fronds 2 celled thick, sub-cartilaginous at the base and membranous at the upper parts; cells in transverse section rectangular or vertically elongated, 1.5 – 2 times longer than broad with cup-shaped chloroplast.
Use	:	Used as food and animal feed in China ("Oyster vegetable"). This is one of the developing aquaculture crops used for food and pharmaceutical substances

Ulvales, Ulvaceae

Distribution
Dwarka: N 22° 14.61'; E 68° 57.28'
Okha: N 22 28.41'; E 69° 04.16'
Porbandar: N 21° 37.92'; E 69° 36.48'
Veraval: N 20° 54.87'; E 70° 20.83'

Ulva fasciata **Delile**

Thallus form and type	:	Leafy and ribbon-like
Substratum	:	Rocks or any other hard substrata in intertidal area
Occurrence	:	Mid littoral zone and tide pools
Season	:	August–April
Abundance	:	Dense during monsoon or rainy periods
Description	:	Plants yellow to dark green in colour, up to 40cm long and divided into a number of ribbon shaped 1 – 3cm broad lobes (Fig. a); blades irregularly lobed or some times divided into ligulate or linear lobes, irregularly dichotomously branched (Fig. b) with entire to ruffled and crenate margins; margins turn whitish when releasing gametes or zoospores; thallus 2 cell thick, cells rectangular or quadratic in cross section.
Remark	:	*Ulva* species are generally used as feed for aquaculture organisms and poultry animals

Ulvales, Ulvaceae

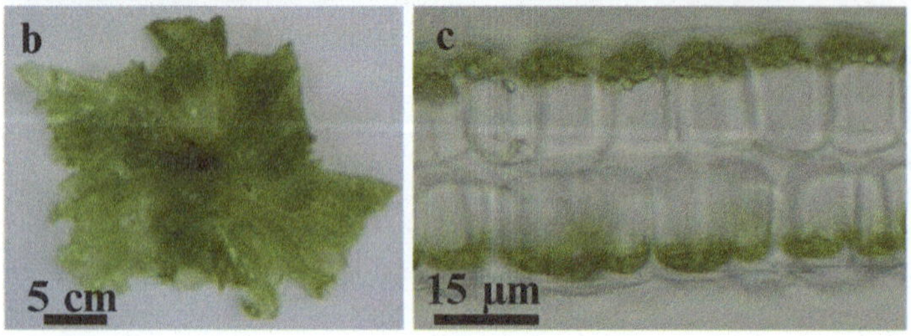

Distribution
Gopnath: N 21° 12.52'; E 72° 06.53'
Dwarka: N 22° 14.61'; E 68° 57.28'
Okha: N 22° 28.33'; E 69° 04.53'
Porbandar: N 21° 37.92'; E 69° 36.48'
Veraval: N 20° 54.87'; E 70° 20.83'

Ulva lactuca **Linnaeus**

Thallus form and type	:	Leafy and membranous
Substratum	:	Rocks/corals
Occurrence	:	Mid littoral zone and tide pools
Season	:	August–April
Abundance	:	Dense
Description	:	Plants soft, leafy, membranous, light to bright green in colour; thallus broad with undulated margins imparting lettuce type morphology (Figs. a, b); surface cells irregularly arranged; thallus distromatic (Fig. c), 40 µm thick, cells 20 µm wide and 15 µm in length and cells with rounded angles.
Use	:	It is used as a traditional medicine in China and commonly called as sea lettuce. The dry powder of this alga used in larval and poultry feed formulations in some countries. Japan alone harvests about 1,500 t (dry wt) from wild stocks. In India, some efforts are being made to develop new applications for liquid extracts of this algae as culture medium supplements for phytoplankton mass culture

Ulvales, Ulvaceae

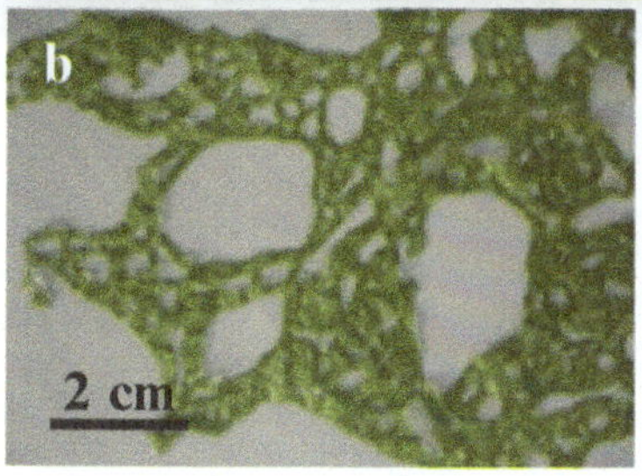

Distribution
Okha: N 22° 27.04'; E 69° 03.58'
Nava Bandar: N 22° 45.33'; E 71° 75.23'

Ulva reticulata **Forsskål**

Thallus form and type	:	Typically perforated leafy thallus
Substratum	:	Rocks or as epiphyte
Occurrence	:	Lower mid littoral zone, seawater streams, tide pools
Season	:	August–April
Abundance	:	Moderate in selective habitats such as streams
Description	:	Plants growing separately or some times in association with other algae, light to dark green in colour, net like or reticulate, membranous, 10 – 20 cm broad with a number of lacunae (Fig. a); lacunae oval, circular, oblong or rectangular (Fig. b), divide the lamina into distinct laciniae with microscopic serrations on the edges of thallus and the holes; thallus 2 cell thick, cells elongated. Plants form dense populations in intertidal pools and streams having domestic sewage outlets.
Use	:	It is used as a substitute for *Ulva lactuca* in traditional medicine in China and Vietnam. There are reports in the literature describing its use in human food in some countries

Ulvales, Ulvaceae

Distribution
Gopnath: N 21° 12.52'; E 72° 06.53'
Porbandar: N 21° 37.92'; E 69° 36.48'

***Ulva rigida* C. Agardh**

Thallus form and type	:	Leafy and leathery
Substratum	:	Intertidal rocks, calcareous stones
Occurrence	:	Lower mid littoral and intertidal pools, sheltered rocks
Season	:	November–April
Abundance	:	Moderate
Description	:	Plants dark green in colour, attached by a distinct holdfast; fronds thick, rigid, 10 – 15 cm tall (Fig. a) with a distinct solid stipe below and lamina above (Fig. b); lamina broadly orbicular or deeply lobed; fronds 2 cells thick; cells longer than broad, palisade-like cells; cell walls thicker than all other species of *Ulva*.
Use	:	There are some records indicating the use of different species of *Ulva* in different human food preparations, animal feed formulations and in medicines. Recently, many researchers considering the use of this biomass for energy needs due to their widespread distribution and abundance in nature

Acrosiphoniales, Acrosiphoniaceae

a

b

Distribution
Dwarka: N 22° 14.61'; E 68° 57.28'
Kotda: N 20° 42.38'; E 70° 52.17'
Nava Bandar: N 22° 45.33'; E 71° 65.23'
Okha: N 22° 28.46'; E 69° 04.34'
Porbandar: N 21° 37.92'; E 69° 36.48'
Veraval: N 20° 54.87'; E 70° 20.83'

Acrosiphonia orientalis **(J. Agardh) P. Silva**

Thallus form and type	:	Remiform and bushy
Substratum	:	Rocks and calcareous stones
Occurrence	:	Mid littoral zone and infra littoral fringe, some times as epiphyte on *Sargassum* and other macroalgae
Season	:	November–April
Abundance	:	Moderate
Description	:	Plants dark green in colour, up to 6 cm tall, bushy and closely branched like corymbose tufts, filamentous (Figs. a, b); filaments pseudo-dichotomously alternately branched from the upper parts or lower half of the branches; cells subcylindrical tapering below with thick walls; the characteristics downwardly growing rhizoids arising from the lower ends of the cells.

Cladophorales, Cladophoraceae

Distribution
Adri: N 20° 57.58'; E 70° 16.76'
Kotda: N 20° 42.38'; E 70° 52.17'
Nava Bandar: N 22° 45.33'; E 71° 65.23'
Okha: N 22° 28.48'; E 69° 04.03'

Chaetomorpha antennina **(Bory) Kützing**

Thallus form and type	:	Filamentous and brush-like tufts in flocks
Substratum	:	Rocks, calcareous stones
Occurrence	:	Lower mid littoral zone and infra littoral fringe, exposed rocks
Season	:	November–April
Abundance	:	Moderate
Description	:	Plants dark green in colour, grows gregariously in tufts; filaments erect, unbranched, cylindrical or barrel-shaped (Figs. a, b), up to 10 cm tall; cells 200 – 250 μm at the upper parts, 2 – 4 times longer than broad with a thick cell wall; chloroplasts reticulate or free, but close together, filaments attached by irregularly branched rhizoidal basal cell; cell walls of the basal cells thick with annular constrictions.
Use	:	Used as raw material for paper manufacturing industries in some countries

Cladophorales, Cladophoraceae

Distribution
Adri: N 20° 57.58'; E 70° 16.76'
Okha: N 22° 28.44'; E 69° 04.07'
Porbandar: N 21° 37.92'; E 69° 36.48'
Veraval: N 20° 54.87'; E 70° 20.83'

Chaetomorpha crassa **(C. Agardh) Kützing**

Thallus form and type	:	Filamentous and entangled mass
Substratum	:	Free floating or as entangled clump on seaweed beds
Occurrence	:	Mid littoral zone, tangles on larger algar
Season	:	November–April
Abundance	:	Scanty
Description	:	Plants usually grow as entangled form with other seaweeds (Fig. a); thallus filamentous; bright green in colour; filaments unbranched with series of cylindrical to barrel shaped cells; cells slightly constricted near the septa, as long as broad or slightly broader, 400 – 650 μm in diameter.
Use	:	There are some reports describing the use of this alga in paper manufacturing industry. The cellulose from this alga could be modified to suit for developing cellulose based membranes for desalination technology

Cladophorales, Cladophoraceae

Distribution
Okha: N 22° 28.41'; E 69° 04.11'
Veraval: N 20° 54.87'; E 70° 20.83'

***Chaetomorpha spiralis* Okamura**

Thallus form type	:	Filamentous and coiled
Substratum	:	Intertidal rocks
Occurrence	:	Lower mid littoral zone, sheltered rocks with coarse sand
Season	:	November–April
Abundance	:	Scanty
Description	:	Plants bluish green in colour, up to 5 cm long, rigid, spirally coiled, attached by rhizoidal basal cells with simple or slightly branched short and blunt rhizoids; cells 400 – 600 µm in diameter and 2 – 3 times longer than broad, moniliform or cylindrical in shape (Figs. a, b).

Cladophorales, Cladophoraceae

Distribution
Nava Bandar: N 22° 45.33'; E 71° 65.23'
Okha: N 22° 28.50'; E 69° 04.54'
Porbandar: N 21° 38.24'; E 69° 35.81'
Veraval: N 20° 54.87'; E 70° 20.83'

Cladophora vagabunda **(Linnaeus) van den Hoek**

Thallus form and type	:	Filamentous and branched tufts
Substratum	:	Rocks, calcareous rocks or as epiphyte on *Sargassum*
Occurrence	:	Mid littoral zone
Season	:	November–April
Abundance	:	Dominant especially during winter months (December–February)
Description	:	Plants light to dark green in colour (Fig. a), forms tufts, 30 – 40 cm in height, densely branched, soft and flaccid, main axis sparingly branched with alternate branches; cells of the main filaments 4 – 6 times as long as broad with highly lamellated cell walls; main filaments and branches bear a short intervals very loosely branched fascicle like tufts of ramuli.

Cladophorales, Cladophoraceae

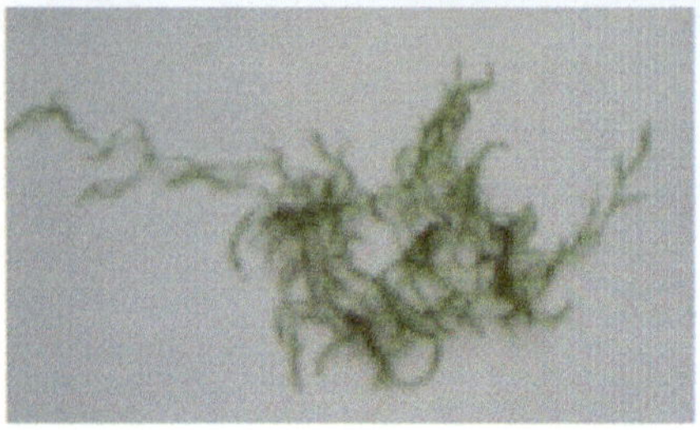

***Cladophora* sp**

Thallus form and type	:	Remiform and filamentous branched
Substratum	:	Rocks/calcareous rocks
Occurrence	:	Mid littoral zone, exposed stones with fine sand
Season	:	November–April
Abundance	:	Scanty
Description	:	Plants grow up to 40 cm high, soft and flaccid, cells of the main filaments 4 – 6 times as long as broad, cell walls are highly lamellated; main filaments and branches bear at short intervals very loosely branched fascile-like tufts of branched ramuli (Figs. a, b).

Siphonocladales, Boodleaceae

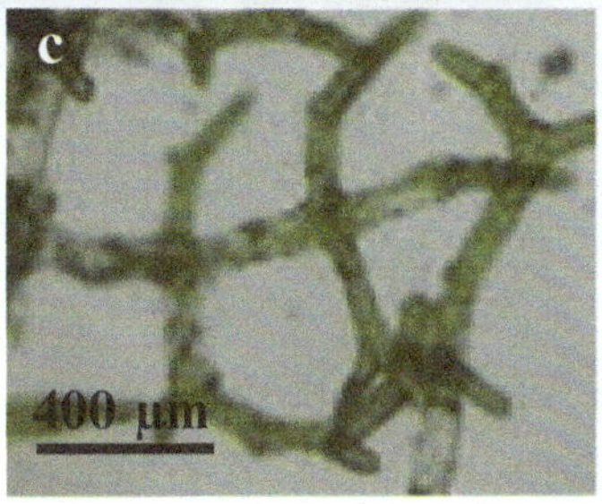

Distribution
Dwarka: N 22° 14.61'; E 68° 57.28'
Okha: N 22° 28.52'; E 69° 04.24'
Porbandar: N 21° 38.92'; E 69° 35.81'
Shivrajpur: N 22° 19.87'; E 68° 56.95'
Veraval: N 20° 54.87'; E 70° 20.83'
Vervala: N 22° 16.87'; E 69° 56.75'

***Boodlea composita* (Harvey) Brand**

Thallus form and type	:	Spongy, branched cushion
Substratum	:	Intertidal rocks, coralline stones
Occurrence	:	Mid littoral zone and infra littoral fringe
Season	:	November–April
Abundance	:	Moderate
Description	:	Plants light or yellowish green in colour, bushy, filamentous, irregularly branched, fragile and spongy, septate, coenocytes (Figs. a, b); branching unilateral or opposite (rarely three or four) arising from the nodes, but soon becoming whorled or irregular and in the lesser divisions altogether irregular, not flabellate (Fig. c); the main filaments 200 – 350 µm in diameter; cells 0.2 mm long, 70 – 100 µm in diameter, devoid of basal cross walls which are acquired later; branches and branchlets attached by teneculae.

Siphonocladales, Boodleaceae

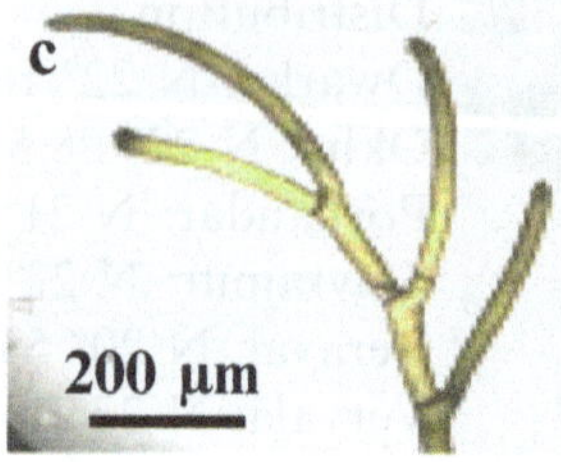

Distribution
Nava Bandar: N 22° 45.33'; E 71° 65.23'
Gopnath: N 21° 12.52'; E 72° 06.53'

Cladophoropsis javanica **(Kützing) P. Silva**

Thallus form and type	:	Spongy, filamentous and prostrate
Substratum	:	Intertidal rocks, coralline stones
Occurrence	:	Upper mid littoral zone
Season	:	November–April
Abundance	:	Scanty
Description	:	Plants yellowish to dark green in colour, occurring as cushions or tufts in the form of entangled uniseriate filaments (Figs. a, b) without a distinct main axis, attached by a characteristic multicellular heptera; erect filaments abundantly laterally branched, branches usually without a basal wall near the bases of the branches (Fig. c); cells large with many nuclei and chloroplasts distributed in a manner of loose net.

Siphonocladales, Boodleaceae

Distribution
Okha: N 22° 28.47'; E 69° 04.23'
Shivrajpur: N 22° 19.87'; E 68° 56.95'
Vervala: N 22° 16.87'; E 69° 56.75'

Struvea anastomosans **(Harvey) Piccone & Grunow ex Piccone**

Thallus form and type	:	Filamentous and fan type
Substratum	:	Intertidal rocks, coralline stones
Occurrence	:	Lower mid littoral zone and infra littoral fringe
Season	:	December–April
Abundance	:	Moderate
Description	:	Plants green to dark green in colour, 3 – 5 cm tall, consists of a long stipe and a circular or oval shaped reticulate blades (Figs. a, b), attached by dichotomously branched rhizoids; stalks un-segmented in the lower portions, segmented above in the form of reticulate flat blades; blades sub-pyramidal to cordate or round, tri-pinnate to quadric-pinnate; branches curved upwards, lateral branches develop tenaculae.

Siphonocladales, Siphonocladaceae

Distribution
Porbandar: N 21° 38.24'; E 69° 35.81'
Veraval: N 20° 54.87'; E 70° 20.83'

Boergesenia forbesii **(Harvey) Feldmann**

Thallus form and type	:	Grape-like vesicles
Substratum	:	Intertidal rocks, corraline stones, dead corals
Occurrence	:	Infra littoral fringe, rock pools
Season	:	November–April
Abundance	:	Scanty
Description	:	Plants yellowish green in colour, 3 – 5 cm tall forming patches on the substratum (Fig. a); vesicles slightly curved, clavate, filled with fluid, club-shaped in younger stage enlarged in the upper part giving grape-like appearance in older stage (Fig. b); basal branched filamentous parts of the thallus septate. Common in intertidal rock pools.
Use	:	The salt prepared from sap can be used in formulations for producing low sodium salt from algal sources

Siphonocladales, Siphonocladaceae

b

Distribution
Adri: N 20° 57.58'; E 70° 16.76'
Kotda: N 20° 42.38'; E 70° 52.17'
Okha: N 22° 28.51'; E 69° 04.37'
Porbandar: N 21° 38.24'; E 69° 35.81'
Sutrapada: N 20° 50.18'; E 70° 28.89'
Veraval: N 20° 54.87'; E 70° 20.83'

Chamaedoris auriculata **Børgesen**

Thallus form and type	:	Bushy with flat heads
Substratum	:	Intertidal rocks, coralline stones
Occurrence	:	Infra littoral zone, tide pools
Season	:	November–April
Abundance	:	Dense in intertidal rock pools
Description	:	Plants dark green in colour, 6 – 8 cm tall, septate coenocytes with long erect stalks, stalks 1 – 3 mm in diameter with annulations (Figs. a, b) lightly calcified with dark green, flattened or eccentric cup-like crown of uniseriate branched uncalcified filaments with polyhedric chloroplast containing one pyrenoid each; crown 2 – 3 cm broad.

Siphonocladales, Siphonocladaceae

b

Distribution
Okha: N 22° 28.43'; E 69° 04.08'
Shivrajpur: N 22° 19.87'; E 68° 56.95'

Dictyosphaeria cavernosa **(Forsskal) Børgesen**

Thallus form and type	:	Hollow, spherical, irregularly lobed or globose
Substratum	:	Intertidal rocks, coralline stones
Occurrence	:	Lower mid littoral zone and infra littoral fringe
Season	:	November–April
Abundance	:	Moderate
Description	:	Plants green in colour, rigid, gregarious, irregularly lobed, hollow, spherical or globose, coenocytic, 2 – 5 cm diameter (Figs. a, b); usually consisting of a single layer of cells formed by segregative cell division resembling large, polygonal, honeycomb-like cells, attached laterally by tenacular cells and attached to the substratum by rhyzoids produced from basal vesicles; polygonal cells 1 – 4 mm broad; tenacular cells 30 – 40 μm long and 20 – 53 μm broad.

Siphonocladales, Valoniaceae

Distribution
Adri: N 20° 57.58'; E 70° 16.76'
Okha: N 22° 28.47'; E 69° 04.39'
Veraval: N 20° 54.87'; E 70° 20.83'

Valonia aegagropila **C. Agardh**

Thallus form and type	:	Compact vesicular masses
Substratum	:	Intertidal rocks, coralline stones
Occurrence	:	Lower mid littoral zone, shallow tide pools
Season	:	August–April
Abundance	:	Moderate
Description	:	Plants dark green in colour, hemispherical in shape, coenocytic, at first attached, later free, eventually forming dense tufts, composed of many short irregularly branched sub-cylindrical straight or arcuate coenocytic vesicles (Figs. a, b); vesicles 1 – 3 mm broad, 5 – 10 mm long; branching from the sides or more usually from the ends of the cells.

Siphonocladales, Valoniaceae

Distribution
Okha: N 22° 28.47'; E 69° 04.37'
Adri: N 20° 57.58'; E 70° 16.76'

***Valonia utricularis* (Roth) C. Agardh**

Thallus form and type	:	Loose vesicle like patches
Substratum	:	Intertidal rocks, coralline stones
Occurrence	:	Lower littoral zone, common in tide pools
Season	:	August–April
Abundance	:	Moderate
Description	:	Plants dark green in colour, spreading as patches among coarser algae, up to 5 cm tall (Figs. a, b); thallus composed of short branched filaments of rather large cylindric–clavate, vesicles or coenocytic cells; vesicles 5 – 15 mm long, 2 – 5 mm diameter; branches arising from the cells cut off at the apices of the vesicles (Fig. c).

Siphonocladales, Valoniaceae

Distribution
Dwarka: N 22° 14.61'; E 68° 57.28'
Kotda: N 20° 42.38'; E 70° 52.17'
Okha: N 22° 28.49'; E 69° 04.39'
Porbandar: N 21° 38.24'; E 69° 35.81'
Veraval: N 20° 54.87'; E 70° 20.83'

Valoniopsis pachynema **(Martens) Børgesen**

Thallus form and type	:	Cushions or spongy patches
Substratum	:	Intertidal rocks, coralline stones
Occurrence	:	Mid littoral zone, exposed rocks
Season	:	September–April
Abundance	:	Moderate
Description	:	Plants dark green in colour, loosely entangled, forming wide cushions of 3 cm thick (Fig. a), 5 – 7 cm diameter with interlaced ascending coenocytic filaments; filaments repeatedly and often sub-corimbosely branched, attached by septate primary and secondary rhizoids; younger branchlets erect, older branches spreading and often arcuate; branches unilateral or palmate, cylindrical, 5 – 7 mm long, 500 – 750 μm diameter, arising from a few small lenticular cells in basal and other parts of the thallus.

Bryopsidales, Bryopsidaceae

Distribution
Nava Bander: N 22° 45.33'; E 71° 65.21'

Bryopsis hypnoides **Lamouroux**

Thallus form and type	:	Remiform with radially arranged pinnules
Substratum	:	Intertidal rocks, coralline stones
Occurrence	:	Lower mid littoral zone
Season	:	November–April
Abundance	:	Moderate
Description	:	Plants yellowish to pale green in colour (Fig. a), prostrate, up to 10 cm tall; erect axes naked below and profusely, radially branched at the upper parts; branches irregular, progressively smaller, without any notable difference between the lesser branches and the slender branchlets; thallus completely coenocytic without septa. Forms dense mats in shaded places of lower intertidal zone.

Bryopsidales, Bryopsidaceae

Distribution
Dwarka: N 22° 14.61'; E 68° 57.28'
Okha: N 22° 28.40'; E 69° 4.50'
Porbandar: N 21° 37.92'; E 69° 36.48'
Rajapara: N 20° 47.46'; E 71° 12.16'
Shivrajpur: N 22° 19.87'; E 68° 56.95'
Veraval: N 20° 54.87'; E 70° 20.83'

Bryopsis pennata **Lamouroux**

Thallus form and type	:	Remiform and feathery
Substratum	:	Intertidal rocks, coralline stones
Occurrence	:	Mid and lower littoral zone, shallow tide pools
Season	:	November–March
Abundance	:	Moderate
Description	:	Plants dark green in colour, sometimes iridescent, up to 14 cm high, siphonous, attached by rhizoidal holdfasts and feather like (Fig. a); erect axes naked below, sparingly branched with pinnately divided branchlets; branchlets distichous, of rather uniform length giving a linear-lanceolate or oblong aspect to the fronds (Fig. b). Forms dense mats in shaded areas below overhanging cliffs.

Bryopsidales, Bryopsidaceae

Distribution
Dwarka: N 22° 14.61'; E 68° 57.28'
Okha: N 22° 28.44'; E 69° 4.50'
Porbandar: N 21° 37.92'; E 69° 36.48'
Rajapara: N 20° 47.46'; E 71° 12.16'
Shivrajpur: N 22° 19.87'; E 68° 56.95'
Sutrapada: N 20° 50.18'; E 70° 28.89'
Veraval: N 20° 54.87'; E 70° 20.83'

Bryopsis plumosa **(Hudson) C. Agardh**

Thallus form and type	:	Remiform and feathery
Substratum	:	Rocks, coralline stones
Occurrence	:	Lower mid littoral zone
Season	:	November–April
Abundance	:	Common
Description	:	Plants light to olive green in colour, growing as tufts or patches, coenocytic, up to 6 cm tall; erect fronds naked below (Fig. a), regularly plumose above with basal rhizomatous portions; branching pinnate and bipinnate with long branches below and short once above giving arrow head appearance (Fig. b).

Bryopsidales, Bryopsidaceae

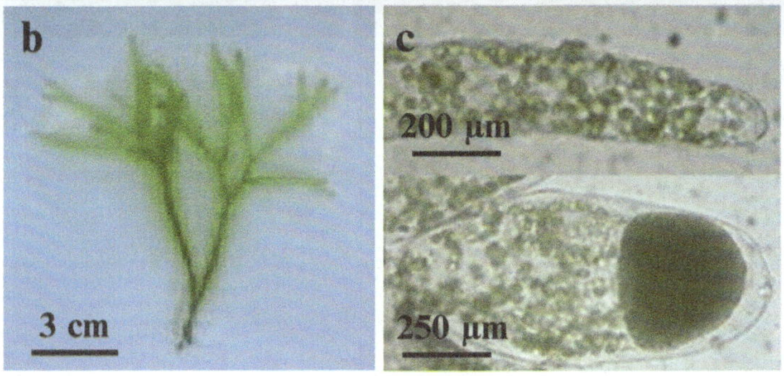

Distribution
Kotda: N 20° 42.38'; E 70° 52.17'
Okha: N 22° 28.41'; E 69° 3.58'
Porbandar: N 21° 37.92'; E 69° 36.48'
Shivrajpur: N 22° 19.87'; E 68° 56.95'

Trichosolen mucronatus **(Børgesen) W. R. Taylor**

Thallus form and type	:	Tubular branched
Substratum	:	Rocks, coralline stones
Occurrence	:	Lower mid littoral zone and on floating structures in the sea
Season	:	November–April
Abundance	:	Rare
Description	:	Plants yellowish green in colour, siphonous, erect, branched thallus, 5 – 10 cm tall, central axis terete, densely covered with cylindrical pinnae from all sides (Figs. a, b); pinnae simple, 160 – 250 µm diameter, 2 – 3 mm long and loosely arranged on the axis; gametangia oblong with mucronate tips (Fig. c). Fronds are of transparent green, glossy and shining and completely coenocytic without septa.

Bryopsidales, Caulerpaceae

Distribution
Porbandar: N 21° 37.92'; E 69° 36.48'

Caulerpa fastigiata **Montagne f.** *delicatula* **Thivy & Visalakshmi**

Thallus form and type	:	Ramiform and creeping
Substratum	:	Intertidal stones buried in sand
Occurrence	:	Lower mid littoral zone
Season	:	September - December
Abundance	:	Moderate
Description	:	Plants dark green in colour, up to 3 cm in height, completely coenocytic, growing as vast mats (Fig. a) with filamentous rhizomes and erect assimilators; assimilators branched, up to third degree, the lateral branches clavate, sometimes crowded, alternate, opposite, 1 – 2 mm long (Fig. b). Unlike other species of this genus, there is no much difference between stoloniferous and erect branches. Forms dense sand binding mats in sheltered areas.

Bryopsidales, Caulerpaceae

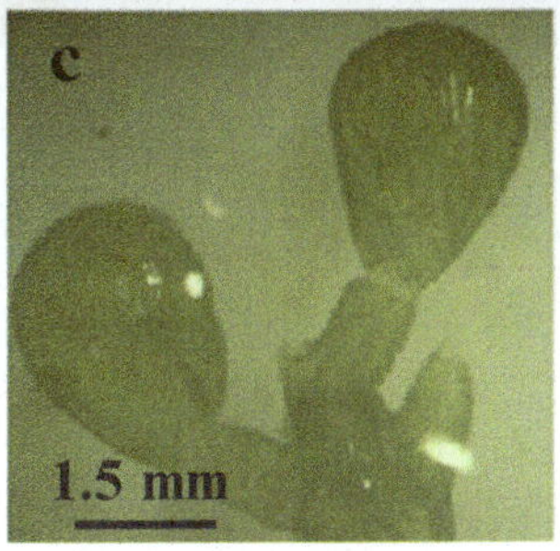

Distribution
Okha: N 22° 28.43'; E 69° 4.17'
Porbandar: N 21° 37.33'; E 69° 36.48'
Samiani Island: N 22° 29.30'; E 69° 05.20'

Caulerpa microphysa **(Weber van Bosse) J. Feldmann**

Thallus form and type	:	Ramiform and creeping
Substratum	:	Intertidal rocks, stones covered with coarse sand
Occurrence	:	Infra littoral fringe, rock crevices
Season	:	November–April
Abundance	:	Moderate
Description	:	Plants bright to dark green in colour, coenocytic, up to 5 cm tall with slender creeping rhizomes and cylindrical erect assimilators (Figs. a, b); ramuli crowded on the assimilators, attached perpendicularly to the axes, sessile or stalked with definite constriction below the globular branchlets (Fig. c). The creeping axes well buried in the sand. It is commonly described as *C. lentillifera*.
Use	:	This alga is used in salad preparations in Japan, Thailand and Philippines. Due to high market demand, it is cultivated in ponds

Bryopsidales, Caulerpaceae

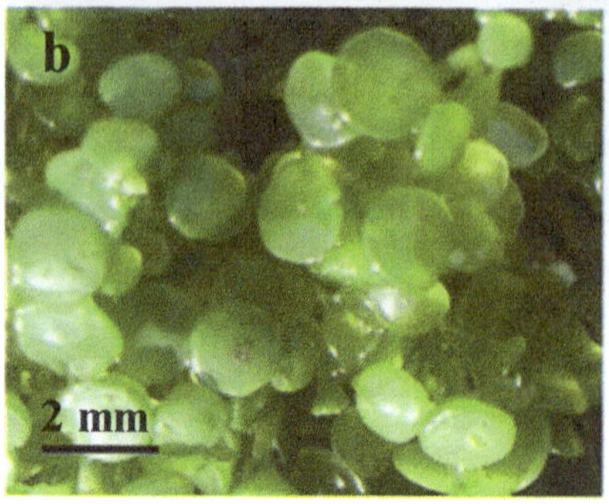

Distribution
Rajpara: N 20° 47.46'; E 71° 12.16'
Samiani Island: N 22° 29.06'; E 69° 05.30'

Caulerpa peltata **Lamouroux**

Thallus form and type	:	Ramiform and creeping
Substratum	:	Intertidal rocks
Occurrence	:	Lower mid littoral zone, exposed rocks
Season	:	November–April
Abundance	:	Scanty
Description	:	Plants bright green in colour, growing as patches, coenocytic with prostrate rhizomes and characteristic short erect assimilators (Fig. a); assimilators 2 – 4 cm tall with radially arranged ramuli; ramuli peltate with slender pedicels or stalks, 1 – 2 mm long ending with a disks of 1 – 2 mm thick and 3 – 8 mm broad (Fig. b). Plants mostly found on sheltered calcareous rocks.

Bryopsidales, Caulerpaceae

Distribution
Dwarka: N 22° 14.61'; E 68° 57.28'
Okha: N 22° 28.44'; E 69° 4.21'
Porbandar: N 21° 37.92'; E 69° 36.48'
Shivrajpur: N 22° 19.87'; E 68° 56.95'
Veraval: N 20° 54.87'; E 70° 20.83'

Caulerpa racemosa **(Forsskål) J. Agardh**

Thallus form and type	:	Ramiform and creeping
Substratum	:	Intertidal rocks, calcareous rocks
Occurrence	:	Lower mid littoral zone, tide pools
Season	:	November–April
Abundance	:	Moderate
Description	:	Plants pale green in colour, growing as patches, coenocytic with prostrate rhizomes and erect assimilators (Fig. a); assimilators often much crowded on the rhizomes, 2 – 5 cm tall with radially and closely arranged ramuli, simple or sparingly forked, covered with clavate to spherical branchlets; stalks of the branchlets short (Fig. b). Most commonly associated with dead corals.

Bryopsidales, Caulerpaceae

Distribution
Okha: N 22° 27.06'; E 69° 4.02'
Porbandar: N 21° 37.92'; E 69° 36.48'
Shivrajpur: N 22° 19.87'; E 68° 56.95'
Veraval: N 20° 54.87'; E 70° 20.83'

Caulerpa racemosa **(Forsskål) J. Agardh v.** ***corynephora*** **(Montagne) Weber-van Bosse**

Thallus form and type	:	Ramiform and creeping
Substratum	:	Intertidal rocks
Occurrence	:	Lower mid littoral zone, common in water streams
Season	:	November–April
Abundance	:	Moderate
Description	:	Plants dark green in colour, occurring in patches, coenocytic, 8 – 10 cm tall with prostrate rhizomes and erect assimilators; assimilators with radially arranged clavate and compressed ramuli.

Bryopsidales, Caulerpaceae

Distribution
Okha: N 22° 28.44'; E 69° 4.20'
Porbandar: N 21° 37.92'; E 69° 36.48'
Shivrajpur: N 22° 19.87'; E 68° 56.95'
Veraval: N 20° 54.87'; E 70° 20.83'

***Caulerpa racemosa* (Forsskål) J. Agardh v. *macrophysa* (Sonder ex Kützing) Taylor**

Thallus form and type	:	Ramiform and creeping
Substratum	:	Intertidal rocks
Occurrence	:	Lower mid littoral zone, common in tidal pools
Season	:	November–April
Abundance	:	Moderate
Description	:	Plants dark green in colour with creeping stolons with short erect assimilators (Fig. a) and rhizoids; assimilators 3 – 5 cm tall, irregularly branched; branchlets arising at 2 – 4 mm intervals, obconical with broad ends, hemispherical to convex, 3 – 5 mm in diameter.

Bryopsidales, Caulerpaceae

a

b

Distribution
Okha: N 22° 28.50'; E 69° 4.54'
Porbandar: N 21° 37.92'; E 69° 36.48'
Veraval: N 20° 54.87'; E 70° 20.83'

***Caulerpa racemosa* v. *occidentalis* (J. Agardh) Børgesen**

Thallus form and type	:	Ramiform and creeping
Substratum	:	Intertidal rocks
Occurrence	:	Lower mid littoral zone, common in tide pools
Season	:	November–April
Abundance	:	Moderate
Description	:	Plants yellowish green in colour, growing as patches, coenocytic; stolons up to 2 mm in diameter with long erect assimilators; assimilators rarely branched, 2 – 10 cm tall or more with radially arranged ramuli; ramuli with long cylindrical stalks and swollen, ovate or globular apices with convex ends (Figs. a, b).

Bryopsidales, Caulerpaceae

Distribution
Okha: N 22° 28.45'; E 69° 4.18'
Porbandar: N 21° 37.92'; E 69° 36.48'
Rajapara: N 20° 47.46'; E 71° 12.16'
Shivrajpur: N 22° 19.87'; E 68° 56.95'
Veraval: N 20° 54.87'; E 70° 20.83'

***Caulerpa racemosa* (Forsskål) J. Agardh var. *turbinata* (J. Agardh) Eubank**

Thallus form and type	:	Ramiform and creeping
Substratum	:	Intertidal rocks
Occurrence	:	Lower mid littoral zone, common in tide pools
Season	:	November–April
Abundance	:	Common
Description	:	Plants bright to dark green in colour, 4 – 6 cm tall, grow as dense patches, siphonous with short erect assimilators; assimilators irregularly branched, covered with turbinate or trumpet shaped ramuli with truncated apices; ramuli 2 mm broad and 5 mm long (Figs. a, b).

Bryopsidales, Caulerpaceae

Distribution
Kotda: N 20° 42.38'; E 70° 52.17'
Okha: N 22° 27.06'; E 69° 4.18'
Porbandar: N 21° 37.92'; E 69° 36.48'
Rajapara: N 20° 47.46'; E 71° 12.16'
Veraval: N 20° 54.87'; E 70° 20.83'

***Caulerpa scalpelliformis* (Brown ex Turner) C. Agardh var. *denticulata* Børgesen**

Thallus form and type	:	Ramiform and creeping
Substratum	:	Intertidal rocks
Occurrence	:	Lower mid littoral zone, rock pools
Season	:	November–April
Abundance	:	Moderate
Description	:	Plants bright green in colour, siphonous with rhizomes and erect foliar assimilators; foliar branches up to 20 cm long, 2 – 3 cm broad, simple with forked or lobed margins, the upper ends of marginal lobes broadly rounded, slightly curved and denticulate (Figs. a, b). Grows commonly on calcareous rocks in tide pools and streams.

Bryopsidales, Caulerpaceae

Distribution
Dwarka: N 22° 14.61'; E 68° 57.28'
Okha: N 22° 28.46'; E 69° 4.24'
Porbandar: N 21° 37.92'; E 69° 36.48'

***Caulerpa scalpelliformis* (Brown ex Turner) C. Agardh f. *dwarkensis* Børgesen**

Thallus form and type	:	Rhizomatous with flat assimilators
Substratum	:	Intertidal rocks, coralline stones
Occurrence	:	Lower mid littoral zone, tide in pools
Season	:	November–April
Abundance	:	Scanty
Description	:	Plants are pale to dark green in colour, siphanaceous, coenocytic with rhizomes and long erect foliar assimilators, simple or branched assimilators; assimilators 20cm long, up to 2cm broad with marginal lobes; lobes mostly straight (Figs. a, b). This alga also grows abundantly in tide pools on sheltered rocks.

Bryopsidales, Caulerpaceae

Distribution
Okha: N 22° 28.42'; E 69° 4.51'
Porbandar: N 21° 37.92'; E 69° 36.48'
Rajapara: N 20° 47.46'; E 71° 12.16'
Veraval: N 20° 54.87'; E 70° 20.83'

***Caulerpa sertularioides* (S. Gmelin) Howe f. *brevipes* (J. Agardh) Svedelius**

Thallus form and type	:	Ramiform and creeping
Substratum	:	Intertidal rocks
Occurrence	:	Lower mid littoral zone, common in tide pools
Season	:	November–April
Abundance	:	Moderate
Description	:	Plants bright to dark green in colour, tubular with delicate, branched stolon and long erect assimilators, plumose or feathery (Fig. a) resembling palm leaves, up to 13 cm tall, 8 – 12 mm broad, rarely branched, branching generally at the base of assimilators; branchlets opposite, incurved above, needle shaped with pointed apices (Fig. b). Common on stones covered with fine sediment.

Bryopsidales, Caulerpaceae

Distribution
Okha: N 22° 28.46'; E 69° 4.51'
Porbandar: N 21° 37.92'; E 69° 36.48'
Samiani Island: N 22° 29.30'; E 69° 05.20'
Veraval: N 20° 54.87'; E 70° 20.83'

***Caulerpa taxifolia* (Vahl) C. Agardh**

Thallus form and type	:	Ramiform and creeping
Substratum	:	Intertidal rocks, calcareous stones
Occurrence	:	Lower mid littoral zone, common in tide pools
Season	:	November–April
Abundance	:	Moderate
Description	:	Plants are yellow to dark green in colour, forming mat-like structures with creeping stolon with erect assimilators (Fig. a); assimilators simple or branched, sometimes with 4 – 5 branches at a time, stalked below, pinnately divided, 4 – 19 cm long and 8 – 13 mm broad; pinnules opposite, compressed, sickle shaped with constricted base, overlapping along edges with pointed apices (Fig. b). It quietly resembles *C. sertularioides* but more robust. It also grows commonly in tide pools on sheltered rocks.

Bryopsidales, Caulerpaceae

a

b

Distribution
Dwarka: N 22° 14.61'; E 68° 57.28'
Okha: N 22° 28.42'; E 69° 4.50'
Porbandar: N 21° 37.92'; E 69° 36.48'
Rajapara: N 20° 47.46'; E 71° 12.16'
Shivrajpur: N 22° 19.87'; E 68° 56.95'
Veraval: N 20° 54.87'; E 70° 20.83'

***Caulerpa veravalensis* Thivy & Chauhan**

Thallus form and type	:	Ramiform and creeping
Substratum	:	Intertidal rocks
Occurrence	:	Lower mid littoral zone, common in tide pools
Season	:	November–April
Abundance	:	Moderate
Description	:	Plants dark green in colour, grows as dense mats on stones covered with fine sand, coenocytic creeping runner with rhizomes and long erect assimilators, assimilators simple or branched; assimilators up to 22 cm tall, 7 – 15 mm broad, compressed, pinnately divided with flat and broad midrib and ramuli (Fig. a); ramuli flat, opposite or alternate, ascending, not curved, 1 – 1.5 mm wide, 5 – 7 mm long with round apices, tips of ramuli rarely bifurcate (Fig. b). It is quite common in tide pools.

Bryopsidales, Caulerpaceae

Distribution
Okha: N 22° 28.39'; E 69° 4.50'
Samiani Island: N 22° 29.30'; E 69° 05.20'

***Caulerpa verticillata* J. Agardh**

Thallus form and type	:	Ramiform and creeping
Substratum	:	Intertidal rocks covered with fine sand
Occurrence	:	Lower mid littoral zone, sheltered rocks
Season	:	November–April
Abundance	:	Moderate
Description	:	Plants deep green in colour, very delicate, stoloniferous with stolons and simple or branched and erect assimilators (Fig. a); assimilators 7 cm tall bearing whorls of cylindrical determinate branchlets (Fig. b); determinate branchlets dichotomously branched 4 – 7 times; ramuli cylindrical, 0.75 – 1.25 mm in length, tapering at the apices; apices mucronate. It grow as dense mats on stones covered with fine sand.

Bryopsidales, Codiaceae

b

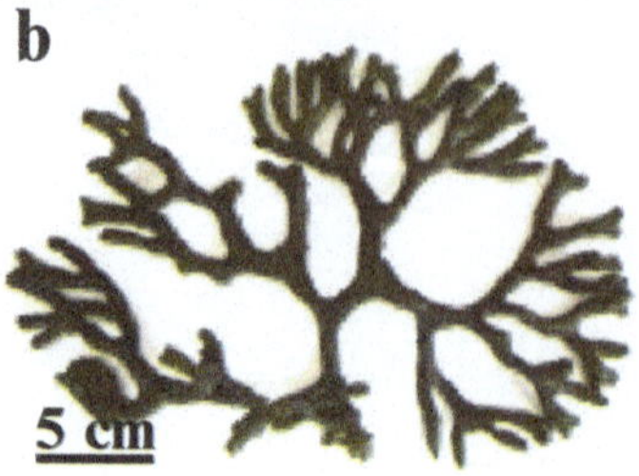

Distribution
Dwarka: N 22° 14.61'; E 68° 57.28'
Okha: N 22° 28.37'; E 69° 3.55'
Porbandar: N 21° 37.92'; E 69° 36.48'
Veraval: N 20° 54.87'; E 70° 20.83'

***Codium decorticatum* (Woodward) Howe**

Thallus form and type	:	Remiform and spongy
Substratum	:	Intertidal rocks, calcareous stones
Occurrence	:	Lower mid littoral zone
Season	:	November–April
Abundance	:	Moderate
Description	:	Plants dark green in colour, 10 – 30 cm tall, 3 – 5 mm broad, bushy, seldom proliferous, attached by basal discs (Fig. a), regularly dichotomous, younger branches of the thallus terate, while older ones frequently compressed and slightly expanded below the furcations (Fig. b); utricles cylindrical or clavate, 150 – 850 μm diameter, apices rounded, truncate or depressed, hair scars variable, numerous when present; gametangia born below the fertile utricles, lance-ovoid, several (up to seven) and pedicellate.
Use	:	*Codium* species are used in food and medicine. It has been reported to contain blood anticoagulant.

Bryopsidales, Codiaceae

b
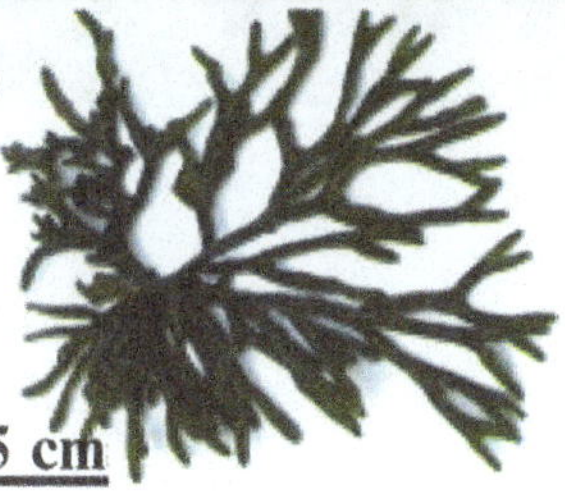

Distribution
Dwarka: N 22° 14.61'; E 68° 57.28'
Okha: N 22° 28.34'; E 69° 3.53'
Porbandar: N 21° 37.92'; E 69° 36.48'
Veraval: N 20° 54.87'; E 70° 20.83'

***Codium dwarkense* Børgesen**

Thallus form and type	:	Remiform and spongy
Substratum	:	Intertidal rocks and calcareous stones
Occurrence	:	Lower mid littoral zone, water streams
Season	:	November–April
Abundance	:	Moderate
Description	:	Plants light green in colour, attached by flat discs with many erect cylindrical axes; erect axes 16 cm or more tall (Fig. a), terete, 2 – 4 mm broad, many times regularly dichotomously divided at a distance of 2 – 3 cm or a little more between the divisions; angles between branches narrow, straight and upwardly directed (Fig. b). Utricles clavate. It also found growing along with *Sargassum* beds.

Bryopsidales, Codiaceae

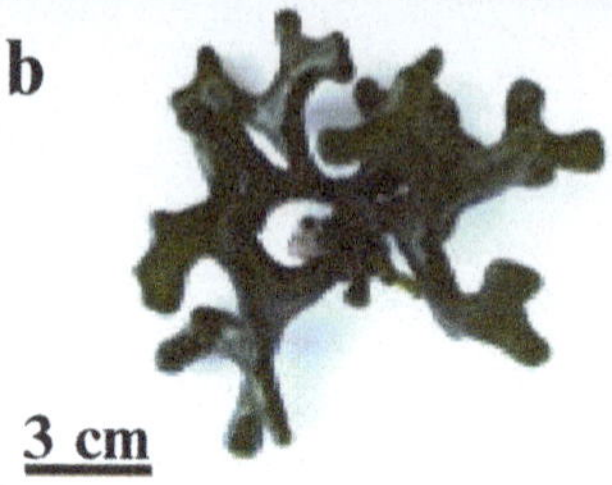

Distribution
Dwarka: N 22° 14.61'; E 68° 57.28'
Okha: N 22° 28.42'; E 69° 3.56'
Samiani Island: N 22° 29.30'; E 69° 05.20'

***Codium geppiorum* O. Schmidt**

Thallus form and type	:	Spongy and procumbent
Substratum	:	Intertidal rocks, calcareous stones
Occurrence	:	Lower mid littoral zone, shallow rock pools
Season	:	November–April
Abundance	:	Scanty
Description	:	Plants dark green in colour, procumbent, attached by holdfast, 3 – 5 cm tall, thallus sub-dichotomously divided and creeping type, branches cylindrical, divergent and compressed (Figs. a, b); utricles abovate to pyriform with rounded tips, 330 – 550 µm long and 100 – 300 µm broad with slight flattened apices; gametangia fusiform, pedicellate, 1 – 2 born on each fertile utricle. Common in tide pools on sheltered rocks.

Bryopsidales, Halimedaceae

b

Distribution
Dwarka: N 22° 14.61'; E 68° 57.28'
Okha: N 22° 28.45'; E 69° 4.20'
Porbandar: N 21° 37.92'; E 69° 36.48'
Shivrajpur: N 22° 19.87'; E 68° 56.95'
Veraval: N 20° 54.87'; E 70° 20.83'

***Halimeda macroloba* Decaisne**

Thallus form and type	:	Articulated and calcified fronds
Substratum	:	Intertidal rocks and calcareous stones
Occurrence	:	Infra littoral fringe and sub littoral zone, tide pools
Season	:	November–April
Abundance	:	Scanty
Description	:	Plants green in colour when not calcified, solitary, erect, 7 – 12 cm long, branched in one plane giving it a flat appearance, attached by a cylindrical holdfast; upper segments discoid, reniform, rounded, 0.5 – 1.5 cm high, 1.5 – 2.0 cm broad, margins thick entire; branching di- to tri-chotomous above and polychotomous at the basal part (Figs. a, b); cortex commonly composed of 3 – 4 layers of utricles, outermost utricles slightly attached when young and separating on decalcification when fertile. It is abundant on dead corals.

Bryopsidales, Halimedaceae

a

b

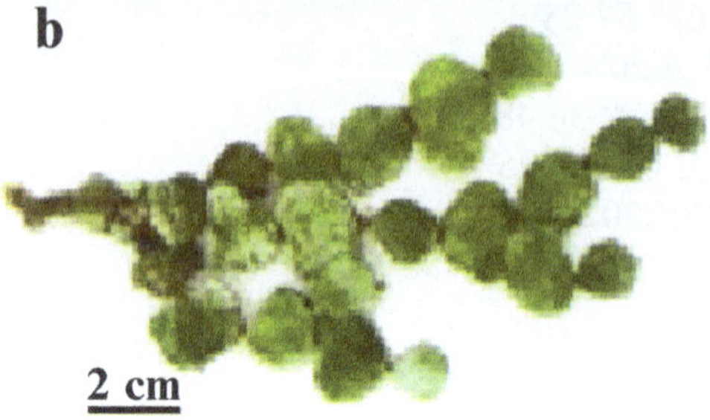

2 cm

Distribution
Dwarka: N 22° 14.61'; E 68° 57.28'
Okha: N 22° 28.45'; E 69° 4.24'
Porbandar: N 21° 37.92'; E 69° 36.48'
Shivrajpur: N 22° 19.87'; E 68° 56.95'
Veraval: N 20° 54.87'; E 70° 20.83'

Halimeda tuna **(Ellis & Solander) Lamouroux**

Thallus form and type	:	Articulated and calcified fronds
Substratum	:	Intertidal rocks, calcareous stones
Occurrence	:	Infra littoral fringe and sub littoral zone
Season	:	November–April
Abundance	:	Scanty
Description	:	Plants dark green in colour, 8 – 10 cm tall, generally tufted, moderately calcified and attached by compressed conical holdfast, branching in one plane, dichotomous or trichotomous upper segments, cuneate remiform discoid, 6 – 13 mm high, 5 – 23 mm broad, margins entire (Figs. a, b); subcortical utricles turbinate to clavate 30 – 110 μm diameter and cortical utricles 25 – 125 μm diameter; medullary filaments entangled, fusing 2 – 3 together at the nodes, dichotomously branched above the fused filaments.

Bryopsidales, Udoteaceae

Distribution
Okha: N 22° 26.38'; E 69° 03.30'

Avrainvillea erecta **(Berkeley) A. et E.S. Gepp.**

Thallus form and type	:	Frondose or brush-like
Substratum	:	Muddy sand
Occurrence	:	Infra littoral zone
Season	:	November–April
Abundance	:	Scanty
Description	:	Plants dark to dull green in colour, erect, up to 8 cm tall, attached by elongated, densely interwoven basal mass of rhizoids with apical fan-shaped sessile and filamentous foliar position (Figs. a, b); foliar portions composed of branched filaments about 4 cm long and 6 cm broad, spongy to somewhat hairy. It commonly found in muddy substratum with still waters.

Bryopsidales, Udoteaceae

Distribution
Okha: N 22° 28.43'; E 69° 4.10'
Porbandar: N 21° 37.92'; E 69° 36.48'
Shivrajpur: N 22° 19.87'; E 68° 56.95'
Veraval: N 20° 54.87'; E 70° 20.83'

***Udotea indica* A. & E. Gepp.**

Thallus form and type	:	Fan like blades
Substratum	:	Intertidal rocks, coralline stones
Occurrence	:	Lower mid littoral zone, tide pools
Season	:	November–April
Abundance	:	Moderate
Description	:	Plants dark green in colour when not calcified, up to 8 cm tall, substantially calcifies (Fig. a) when aged; erect fronds with a tangled rhizoidal base, 2 – 3 cm stalk and terminal fan like or funnel-shaped, rounded flabellate blades (Fig. b), about 5 cm broad; stalk and blade both composed of dichotomously branched filaments, constricted above the forks and occasionally elsewhere.

Dasycladales, Dasycladaceae

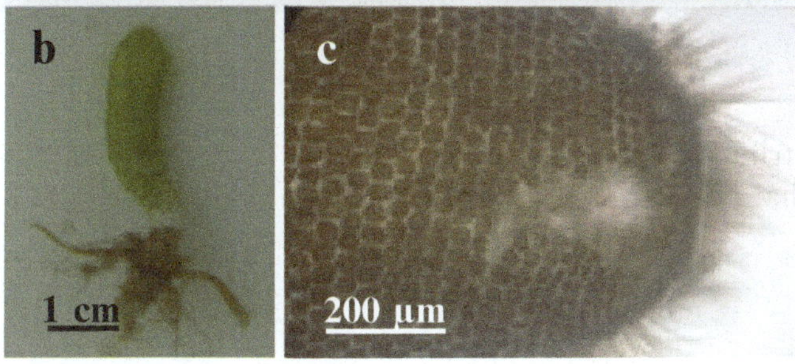

Distribution
Shivrajpur: N 22° 19.87'; E 68° 56.95'

Neomeris annulata **Dickie**

Thallus form and type	:	Tubular
Substratum	:	Intertidal rocks, coralline stones
Occurrence	:	Lower mid littoral zone, tide pools
Season	:	November–April
Abundance	:	Scanty
Description	:	Plants cylindrical and unbranched with bright green apex and calcified whitish basal portions, solitary, often densely gregarious, 2 – 2.5 cm tall (Figs. a, b); axes covered with small whorls of short, compound, 200 – 300 µm long, deciduous branchlets (Fig. c); the expanded capitate ends of the outer cells polyhedral, forming a fairly even faceted and annulated surface. Usually found on rocks or dead corals covered with coarse sand in tide pools.

Dasycladales, Polyphysaceae

Distribution
Okha: N 22° 26.38'; E 69° 03.30'

Acetabularia calyculus **Lamouroux**

Thallus form and type	:	Long stalked cup
Substratum	:	Intertidal rocks, stones, pebbles, dead muddy corals
Occurrence	:	Lower mid littoral zone, muddy substrata
Season	:	November–April
Abundance	:	Scanty
Description	:	Plants light green in colour, slightly calcified stipe, 1.5 – 3.0 cm tall, attached by lobed holdfasts (Figs. a, b); stipes slender with cup shaped discs and branched hairs at the apices, discs with 22 – 30 rays, lightly connected to each other; rays cylindrically clavate with rounded or markedly notched edges; cysts about 80 in each rays, 160 µm in diameter (Fig. c). This alga is found on dead shells and corals that have been covered with mud or fine sand.

Phaeophyta

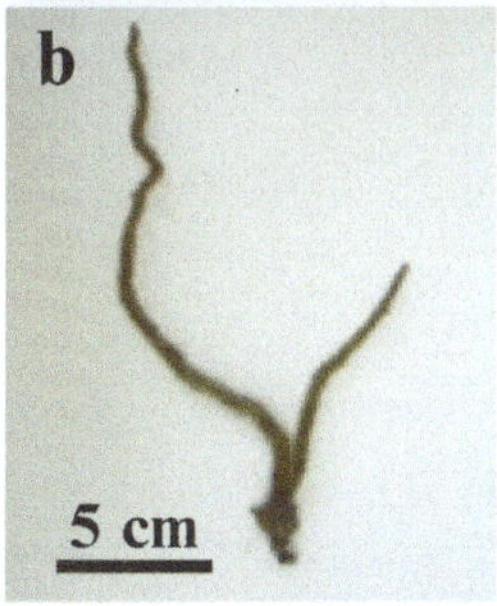

Distribution
Adri: N 20° 57.58'; E 70° 16.76'
Okha: N 22° 28.45'; E 69° 04.23'
Veraval: N 20° 54.87'; E 70° 20.83'

Levringia boergesenii **Kylin**

Thallus form and type	:	Remiform, elastic and hairy
Substratum	:	Intertidal rocks, coralline stones
Occurrence	:	Lower mid littoral zone, tide pools, wave exposed rocks
Season	:	December–April
Abundance	:	Scanty
Description	:	Plants dark to blackish brown in colour, 20 – 30 cm tall, elastic, gelatinous, irregularly branched, growing as loose festoons and attached by basal minute discs (Fig. a); branches present only on the lower portion of the main axis, upper portion poorly branched (Fig. b); thallus multi-axial with a central medulla and a peripheral cortex; long assimilatory hairs present at the apical region.

B. Jha et al., *Developments in Applied Phycology*,
DOI 10.1007/978-90-481-2488-6_2, © Springer Science + Business Media B.V. 2009

Ectocarpales, Acinetosporaceae

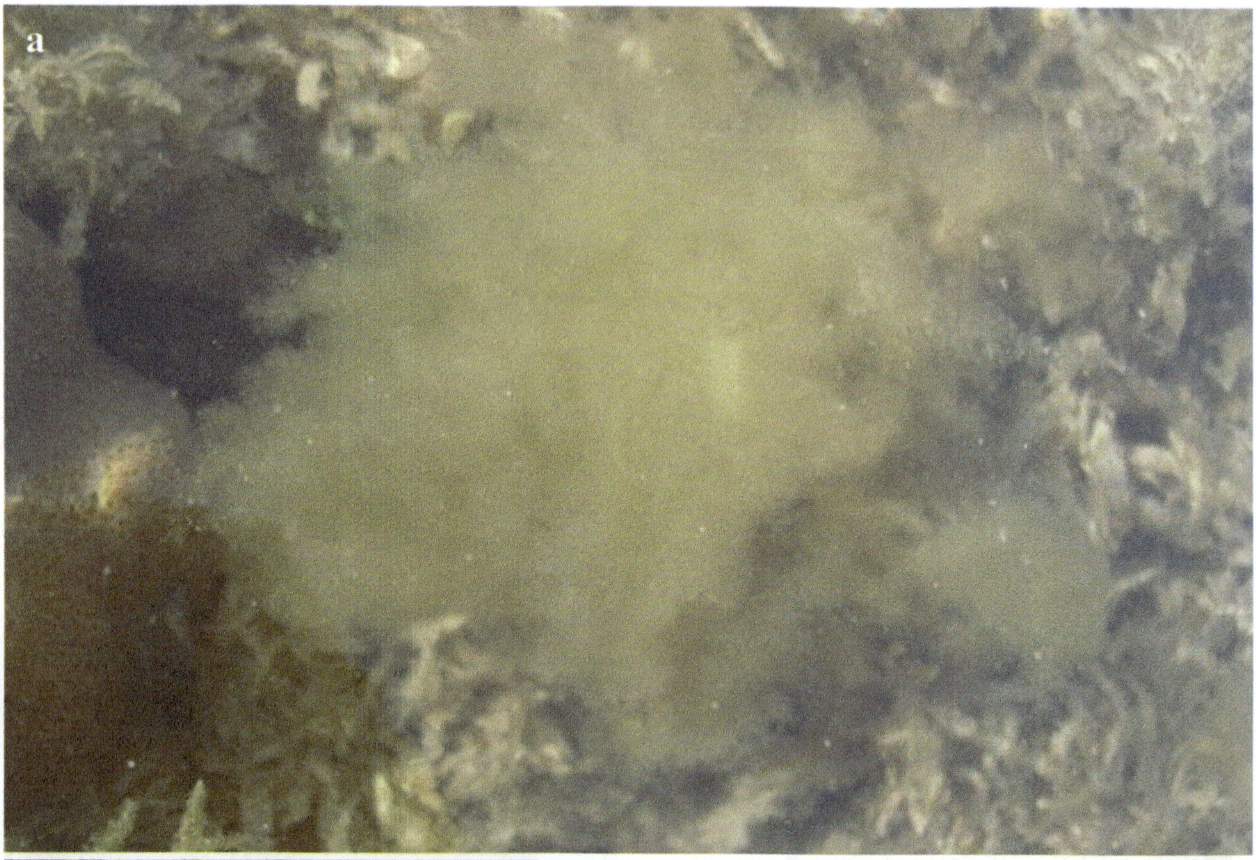

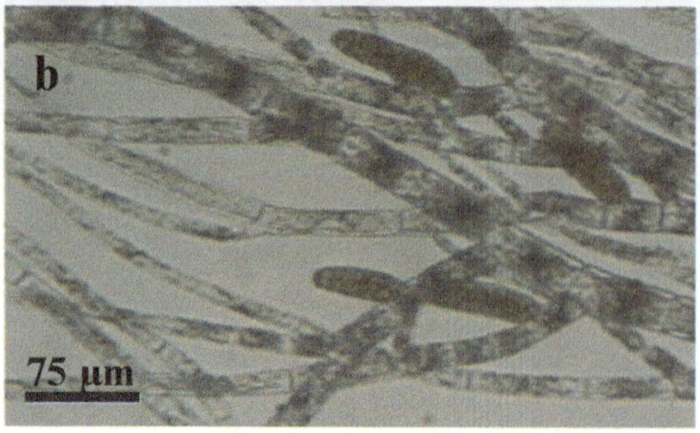

Distribution
Shivrajpur: N 22° 19.87'; E 68° 56.95'

Hincksia mitchelliae **(Harvey) Silva**

Thallus form and type	:	Cloudy tufts
Substratum	:	Intertidal rocks
Occurrence	:	Mid littoral zone and sub littoral zone
Season	:	July–November
Abundance	:	Moderate
Description	:	Plants light brown in colour, mostly epiphytic, growing as tufts, 5 – 10cm tall, attached by means of branched rhizoids from the basal parts of the filaments, erect filaments profusely, monopodialy and irregularly branched, gradually tapering into hair like structures, cells rectangular to barrel shaped with band shaped chloroplast (Fig. a); plurilocular gametangia cylindrically elongated, sessile, rarely stalked, with obtuse apices (Fig. b).

Dictyotales, Dictyotaceae

Distribution
Okha: N 22° 28.43'; E 69° 03.57'
Shivrajpur: N 22° 19.87'; E 68° 56.95'

Dictyopteris acrostichoides **(J. Agardh) Bornet**

Thallus form and type	:	Frondose and ribbon type
Substratum	:	Inter tidal rocks, gravel, coralline stones
Occurrence	:	Lower mid littoral zone
Season	:	November–April
Abundance	:	Scanty
Description	:	Plants dark brown in colour, up to 15 cm tall, arising from discoid holdfasts (Fig. a); fronds dichotomously branched with thin wings and prominent mid rib, 0.4 – 0.5 cm broad, margins entire, apices round (Fig. b); thallus many cells thick in the midrib region two layered in the wings; reproductive organs and hairs forming a narrow belt on both sides of the midrib.

Dictyotales, Dictyotaceae

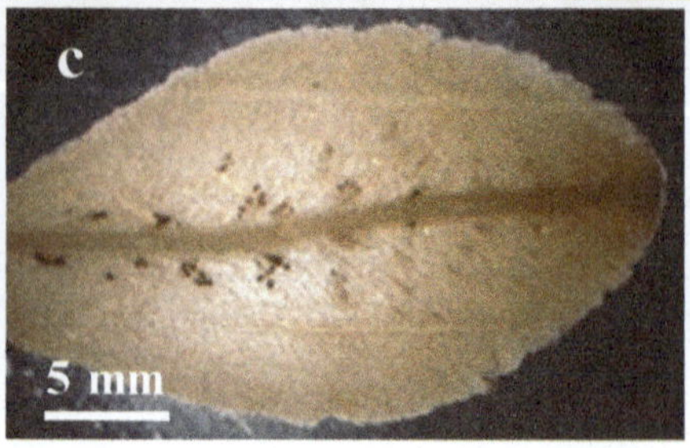

Distribution
Adri: N 20° 57.58'; E 70° 16.76'
Dwarka: N 22° 14.61'; E 68° 57.28'
Okha: N 22° 28.42'; E 69° 03.58'
Porbandar: N 21° 38.24'; E 69° 35.81'
Shivrajpur: N 22° 19.87'; E 68° 56.95'
Veraval: N 20° 54.87'; E 70° 20.83'

Dictyopteris australis **(Sonder) Askenasy**

Thallus form and type	:	Frondose and ribbon type
Substratum	:	Intertidal rocks, gravel and coralline stones
Occurrence	:	Lower mid littoral zone
Season	:	November–April
Abundance	:	Moderate
Description	:	Plants dark brown in colour, up to 25 cm tall, 0.5 – 1.5 cm broad, arising from a small cuneate disc (Figs. a, b); thallus ribbon shaped, membranous, irregularly di- or tri-chotomously branched with a prominent midrib throughout the plant; small veins arising from the midrib, groups of hairs present on both the side of the thallus; proliferations arises from the midrib; margins of the branches wavy, irregularly dentate; tetrasporangia forming recurved arched groups (Fig. c).

Dictyotales, Dictyotaceae

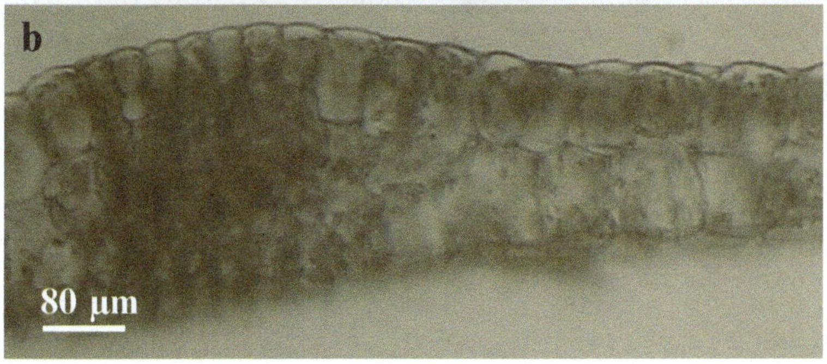

Distribution
Shivrajpur: N 22° 19.87'; E 68° 56.95'

Dictyopteris delicatula **Lamouroux**

Thallus form and type	:	Frondose and ribbon type
Substratum	:	Intertidal rocks, coralline stones
Occurrence	:	Lower mid littoral zone, tidal streams
Season	:	November–April
Abundance	:	Scanty
Description	:	Plants light brown in colour, 3 – 5 cm tall, attached by a minute disc; thallus dichotomously and irregularly branched with entire margins (Fig. a); apices round and midrib more prominent at the lower parts of the thallus, 4 – 6 layered in the mid rib region and two layered in the wings (Fig. b); reproductive organs and hairs forming a narrow belt on both sides of the midrib.

Dictyotales, Dictyotaceae

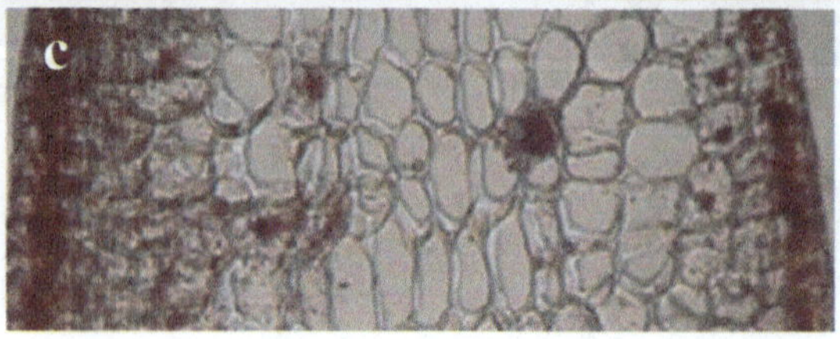

Distribution
Okha: N 22° 28.50'; E 69° 04.56'

Dictyopteris serrata **(Areschoug) Hyot**

Thallus form and type	:	Leafy and blade type
Substratum	:	Intertidal rocks, coralline substrata
Occurrence	:	Subtidal
Season	:	July–November
Abundance	:	Scanty
Description	:	Plants light to dark brown in colour, 22 cm tall, attached by rhizoidal holdfast, dichotomously branched; fronds 3 – 4 cm broad with prominent midrib, wings and lateral veins (Fig. a); lateral veins alternately arising from the midrib, margins with prominent regularly dentate teeth (Fig. b); paraphyses and antheridial sori scattered in the wings; thallus 20 celled thick in midrib region and 4 – 5 cells thick at the margin (Fig. c).

Dictyotales, Dictyotaceae

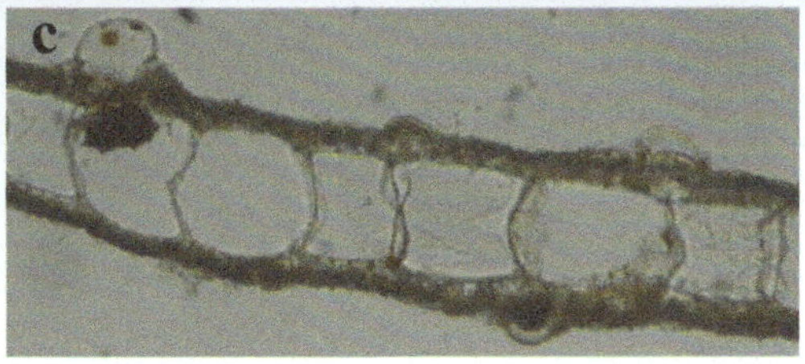

Distribution
Adri: N 20° 57.58'; E 70° 16.76'

Dictyota bartayresiana **Lamouroux**

Thallus form and type	:	Ramiform and foliose or ribbon shaped
Substratum	:	Intertidal rocks, coralline stones
Occurrence	:	Mid littoral and lower littoral areas
Season	:	November–April
Abundance	:	Moderate
Description	:	Plants erect, branched, dark brown in colour, about 10 cm tall (Fig. a, b), branches are strap-like, alternatively to dichotomously divided, branched at narrow angles, gradually decreasing in width to the pointed apex, aculeate tips, blades have spines on their surface, blades 4 – 5 mm wide with a dotted appearance, blade surface rough. Thallus is 3 cell thick (Fig. c). It mostly grows in tide pools and some times as epiphyte on larger brown algae.

Dictyotales, Dictyotaceae

Distribution
Kotda: N 20° 42.38'; E 70° 52.17'
Okha: N 22° 28.54'; E 69° 04.38'
Shivrajpur: N 22° 19.87'; E 68° 56.95'

Dictyota cervicornis **Kützing**

Thallus form and type	:	Frondose, narrow and ribbon type
Substratum	:	Intertidal rocks, coralline substrata
Occurrence	:	Lower mid littoral zone
Season	:	November–April
Abundance	:	Scanty
Description	:	Plants yellowish brown in colour, 10 – 25 cm tall, erect, twisted thallus common, membranous, entangled, irregularly dichotomous, attached by small holdfast; fronds 0.8 – 1.5 cm broad at the base, 2 – 4 mm towards apices; margins entire with rounded and acute apices (Figs. a, b). it is quite common in tide pools with *Sargassum* beds.

Dictyotales, Dictyotaceae

Distribution
Dwarka: N 22° 14.61'; E 68° 57'. 28'
Kotada: N 20° 42.38'; E 70° 52.17'
Shivrajpur: N 22° 19.87'; E 68° 56.95'

Dictyota ciliolata **Kützing**

Thallus form and type	:	Frondose and bushy
Substratum	:	Intertidal rocks, coralline stones
Occurrence	:	Mid littoral zone, tide pools
Season	:	November–April
Abundance	:	Moderate
Description	:	Plants yellowish brown in colour, up to 15 – 20cm in height (Fig. a, b), membranous, ribbon like, upper parts regularly dichotomously branched and attached to the substratum by a small cuneate disc. Branches usually of uniform breadth, forking angles between the branches acute or rounded, apices acute or subacute. Margin of the branches subentire or regularly dentate, possessing small ascending projections. Tetrasporangia forming scattered group. Oogonia and antheridia both distributed all over the surface of the thallus.

Dictyotales, Dictyotaceae

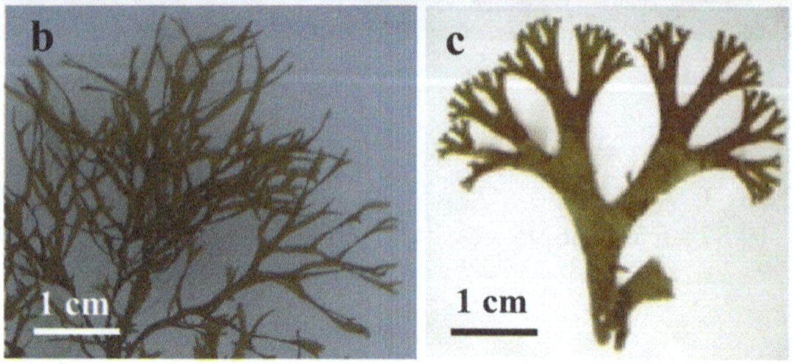

Distribution
Okha: N 22° 28.51'; E 69° 04.51'
Shivrajpur: N 22° 19.87'; E 68° 56.95'
Veraval: N 20° 54.87'; E 70° 20.83'

Dictyota dichotoma **(Hudson) Lamouroux**

Thallus form and type	:	Frondose and bushy
Substratum	:	Intertidal rocks, coralline stones
Occurrence	:	Lower mid littoral zone, tide pools
Season	:	November–April
Abundance	:	Dense
Description	:	Plants yellowish brown in colour, 12 cm or more in height, ribbon like, upper parts regularly dichotomously branched, attached by a small cuneate disc (Figs. a, b); branches usually of uniform breadth with variations in the width of the fronds (Figs. b, c), forking angles of 15 – 45°, axes narrow, margins entire, apices rounded, obtuse or sometimes acute; tetrasporangia single or double, scattered on both sides of the thallus.

Dictyotales, Dictyotaceae

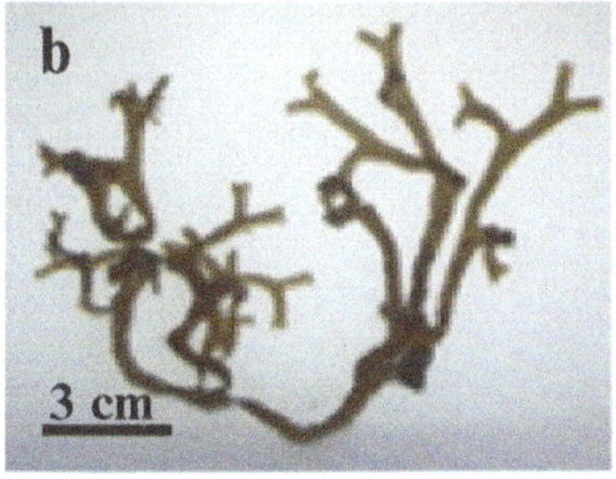

Distribution
Kotda: N 20° 42.38'; E 70° 52.17'
Okha: N 22° 28.54'; E 69° 04.40'
Shivrajpur: N 22° 19.87'; E 68° 56.95'

Dictyota pinnatifida **Kützing**

Thallus form and type	:	Frondose and ribbon type
Substratum	:	Intertidal rocks, coralline stones
Occurrence	:	Lower mid littoral zone
Season	:	November–April
Abundance	:	Scanty
Description	:	Plants light brown in colour, up to 10 cm in height, attached by cuneate discs; thallus flat, irregularly dichotomously to alternately branched, 2 – 3 mm broad; margins entire, angles between branches about 80° with round apices (Figs. a, b).

Dictyotales, Dictyotaceae

Distribution
Adri: N 20° 57.58'; E 70° 16.76'
Okha: N 22° 28.43'; E 69° 04.10'
Porbandar: N 21° 38.24'; E 69° 35.81'
Shivrajpur: N 22° 19.87'; E 68° 56.95'
Veraval: N 20° 54.87'; E 70° 20.83'

***Lobophora variegata* (Lamouroux) Womersley ex Oliveria**

Thallus form and type	:	Crustose
Substratum	:	Intertidal rocks, calcareous stones
Occurrence	:	Infra littoral fringe, rock pools
Season	:	November–April
Abundance	:	Dominant
Description	:	Plants dark brown in colour, 10 – 15 cm broad, prostrate, overlapping in clusters, upper portions entire, lobate with variegated markings in concentric zones; thalli circular to fan shaped or sub-orbicular when young (Figs. a, b) and attached by rhizoids arising from the basal parts of the fronds, composed of a central layer of rectangular medullary cells, 3 – 4 layers of sub-cortical cells and cortical layer of smaller cells.

Dictyotales, Dictyotaceae

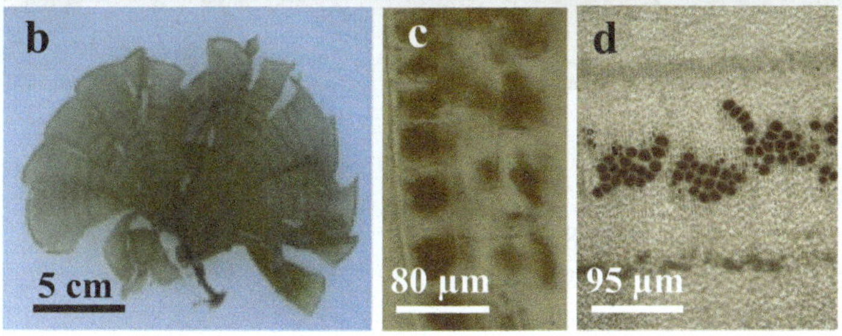

Distribution
Okha: N 22° 28.42'; E 69° 04.16'
Porbandar: N 21° 38.24'; E 69° 35.81'
Shivrajpur: N 22° 19.87'; E 68° 56.95'

Padina boergesenii **Allender & Kraft**

Thallus form and type	:	Frondose and fan shaped
Substratum	:	Intertidal rocks
Occurrence	:	Mid littoral zone, tide pools
Season	:	November–April
Abundance	:	Dominant
Description	:	Plants light to dark brown in colour, up to 15 cm tall, 12 cm in breadth, basal portion forming rhizomatous discs; thallus divided into broad lobes (Figs. a, b), three cells thick (Fig. c), hairs developing all over the thallus as concentric zones, the reproductive structures occurring in the central parts of the thallus between the two hairy zones (Fig. d). It grows as dense mats in large rock pools in intertidal zone.

Dictyotales, Dictyotaceae

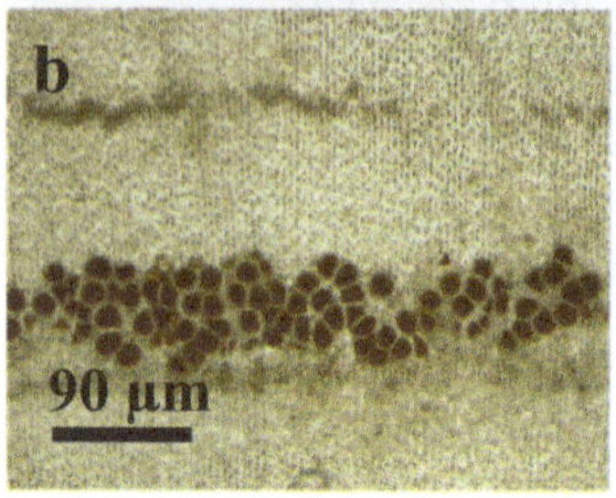

Distribution
Okha: N 22° 28.43'; E 69° 04.19'
Porbandar: N 21° 38.24'; E 69° 35.81'
Shivrajpur: N 22° 19.87'; E 68° 56.95'

Padina boryana **Thivy**

Thallus form and type	:	Frondose and fan shaped
Substratum	:	Intertidal rocks, coralline stones
Occurrence	:	Mid littoral zone, tide pools with coarse sand
Season	:	November–April
Abundance	:	Dominant
Description	:	Plants light brown in colour, 8 – 10 cm tall, 2 – 4 cm broad, attached by rhizomatous discs, fronds fan like, deeply lobed, margin inrolled (Fig. a); concentric rows of hairs and tetrasporangia alternately arranged and sporangia formed just above the rows of hairs (Fig. b). It grows as patches in tide pools.

Dictyotales, Dictyotaceae

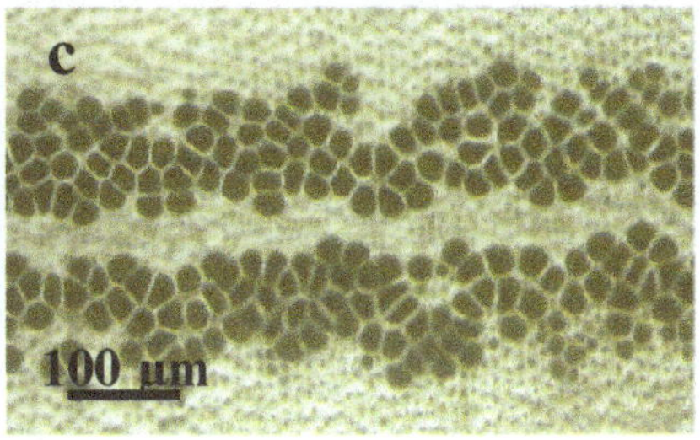

Distribution
Okha: N 22° 28.41'; E 69° 04.13'
Porbandar: N 21° 38.24'; E 69° 35.81'
Shivrajpur: N 22° 19.87'; E 68° 56.95'

Padina tetrastromatica **Hauck**

Thallus form and type	:	Frondose and fan shaped
Substratum	:	Intertidal rocks
Occurrence	:	Mid littoral zone, tide pools with coarse sand
Season	:	November–April
Abundance	:	Dominant
Description	:	Plants yellowish brown in colour, 10 – 15 cm tall with a number of fan shaped blades developing from rhizomatous discs; thallus irregularly cleft into narrow lobes in mature plants (Figs. a, b); mature fronds composed of four layers of cells, reproductive bodies forming as concentric rows on both sides of the rows of hairs (Fig. c). It prefer areas exposed to waves.

Dictyotales, Dictyotaceae

Distribution
Adri: N 20° 57.58'; E 70° 16.76'
Okha: N 22° 29.00'; E 69° 04.39'
Porbandar: N 21° 38.24'; E 69° 35.81'
Shivrajpur: N 22° 19.87'; E 68° 56.95'
Veraval: N 20° 54.87'; E 70° 20.83'

Spatoglossum asperum **J. Agardh**

Thallus form and type	:	Frondose and ribbon type
Substratum	:	Intertidal rocks
Occurrence	:	Infra littoral fringe, tide pools
Season	:	November–April
Abundance	:	Dominant
Description	:	Plants dark brown in colour, 30 – 40 cm tall, arising from rhizomatous compact discs, ribbon-like, irregularly sub-dichotomously or sub-palmately divided; frond segments attenuated at the bases and apices, 1 – 3 cm broad, margins wavy and sometimes dentate with small proliferations (Figs. a, b); thallus with 3 – 4 layers of large, angular cells at the centre and palisade-like, thin walled superficial layers.

Dictyotales, Dictyotaceae

Distribution
Adri: N 20° 57.58'; E 70° 16.76'
Kotda: N 20° 42.38'; E 70° 52.17'
Okha: N 22° 28.54'; E 69° 04.39'
Porbandar: N 21° 38.24'; E 69° 35.81'
Shivrajpur: N 22° 19.87'; E 68° 56.95'
Veraval: N 20° 54.87'; E 70° 20.83'

***Stoechospermum marginatum* (C. Agardh) Kützing**

Thallus form and type	:	Frondose and ribbon type
Substratum	:	Intertidal rocks
Occurrence	:	Infra littoral fringe and sub littoral zone, tide pools
Season	:	November–April
Abundance	:	Dominant
Description	:	Plants yellowish brown in colour, up to 20 cm in height, erect, growing as tufts; thallus ribbon shaped, dichotomously divided, attached by rhizoidal holdfast (Figs. a, b); margins of the branches entire with involute apices; thallus with 4 – 5 celled medulla at the centre and one-celled thick, pigmented peripheral layers; hairs found in large groups, scattered over the surface of the thallus, reproductive bodies occurring as marginal sori.

Scytosiphonales, Scytosiphonaceae

Distribution
Okha: N 22° 28.50'; E 69° 04.30'
Shivrajpur: N 22° 19.87'; E 68° 56.95'
Veraval: N 20° 54.87'; E 70° 20.83'

Colpomenia sinuosa **(Martens ex Roth) Derbes & Solier**

Thallus form and type	:	Globular or lobed vesicles
Substratum	:	Intertidal rocks, epiphytic on other large algae
Occurrence	:	Lower mid littoral zone and infra littoral fringe, wave exposed rocks
Season	:	November–April
Abundance	:	Dominant
Description	:	Plants yellow brown in colour, 5 – 10cm or more in diameter, globular, hollow, forming air-filled and irregularly lobed vesicles with corrugated surface (Figs. a, b); thallus composed of a pigmented layer of epidermis, 5 – 6 layers of large colourless cells below the epidermis, plurilocular sori on the epidermis, associated with paraphyses and hairs.

Scytosiphonales, Scytosiphonaceae

a

b

Distribution
Dwarka: N 22° 14.61'; E 68° 57.28'
Okha: N 22° 28.50'; E 69° 04.30'
Porbandar: N 21° 38.24'; E 69° 35.81'
Shivrajpur: N 22° 19.87'; E 68° 56.95'
Veraval: N 20° 54.87'; E 70° 20.83'
Vervala: N 22° 16.87'; E 69° 56.75'

Hydroclathrus clathratus **(C. Agardh) Howe**

Thallus form and type	:	Vesicular and net like masses
Substratum	:	Intertidal rocks, coralline stones
Occurrence	:	Lower mid littoral zone and infra littoral fringe, wave exposed rocks
Season	:	November–April
Abundance	:	Dominant
Description	:	Plants yellowish to dark brown in colour, up to 12 cm long, 4 – 8 cm broad, vesicular or irregularly ovate, un-perforated when young, hollow and perforated as net like structures (Figs. a, b) when mature; perforations variable in size, rounded, margins around perforations involute; thallus composed of an epidermal layer and a zone of colourless cells below the epidermis; plurilocular sori scattered all over the surface of the thalli, hairs grouped in shallow depression.
Use	:	It is used as food and fertilizer in some countries

Scytosiphonales, Scytosiphonacea

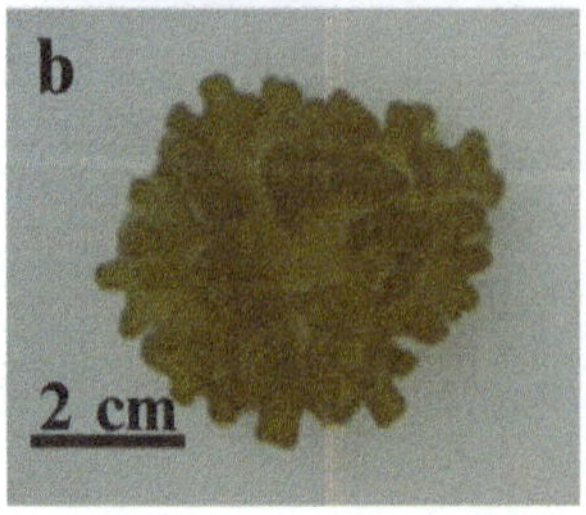

Distribution
Dwarka: N 22° 14.61'; E 68° 57.28'
Okha: N 22° 28.52'; E 69° 04.32'
Porbandar: N 21° 38.24'; E 69° 35.81'
Shivrajpur: N 22° 19.87'; E 68° 56.95'
Veraval: N 20° 54.87'; E 70° 20.83'
Vervala: N 22° 16.87'; E 69° 56.75'

***Iyengaria stellata* (Børgesen) Børgesen**

Thallus form and type	:	Lobed with stellate or antler-like projection
Substratum	:	Intertidal rocks, coralline stones
Occurrence	:	Lower mid littoral zone and infra littoral fringe, wave exposed rocks
Season	:	November–April
Abundance	:	Dominant
Description	:	Plants yellowish to dark brown in colour, 5 – 10 cm tall, young plants are internally solid, smooth, globular, attached by rhizoids arising from the lower epidermal cells (Fig. a); mature plants irregularly lobed with solid, stellate, or antler-like projections (Fig. b); thallus consisting of an epidermal layer and many inner layers of colourless cells; plurilocular sori scatter on the surface, associated with the paraphyses and hairs.

Scytosiphonales, Scytosiphonacea

Distribution
Shivrajpur: N 22° 19.87'; E 68° 56.95'

Rosenvingea intricata **(J. Agardh) Børgesen**

Thallus form and type	:	Branched and tubular tufts
Substratum	:	Intertidal rocks
Occurrence	:	Mid littoral zone, tide pools, wave exposed rocks
Season	:	November–April
Abundance	:	Scanty
Description	:	Plants yellowish brown in colour, 3 – 5 cm tall, growing as intricate tufts with irregularly ramified branches; mature thalli hollow; cells in the thallus larger near the cavity and smaller on the peripheral side and covered with a layer of small polygonal cells containing disc-shaped, roundish or slightly lobed chloroplasts; hairs scattered or in small groups.

Scytosiphonales, Chnoosporaceae

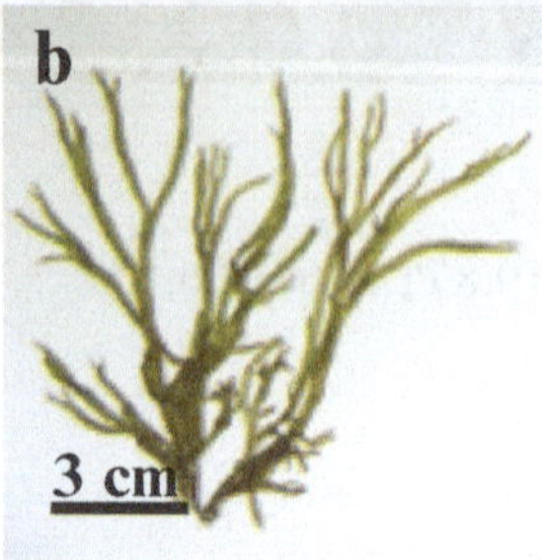

Distribution
Okha: N 22° 28.52'; E 69° 04.23'
Shivrajpur: N 22° 19.87'; E 68° 56.95'

Rosenvingea orientalis **J. Agardh**

Thallus form and type	:	Branched and tubular tufts
Substratum	:	Intertidal rocks
Occurrence	:	Mid littoral zone, tide pools
Season	:	November–April
Abundance	:	Scanty
Description	:	Plants yellowish brown in colour, up to 9 cm tall, tubular tufts, rarely 1 – 5 cm in breadth, tufts irregularly branched with antler-like appearance (Figs. a, b); thallus membranous, cells of surface layer small and subsurface cells are larger; hairs present here and there on the thallus; plurilocular sporangia in small oval groups and scattered all over the thallus.

Fucales, Sargassaceae

Distribution
Dwarka: N 22° 14.61'; E 68° 57.28'
Okha: N 22° 28.43'; E 69° 04.09'
Porbandar: N 21° 38.24'; E 69° 35.81'
Veraval: N 20° 54.87'; E 70° 20.83'

Cystoseira indica **(Thivy & Doshi) Mairh**

Thallus form and type	:	Remiform and bushy
Substratum	:	Intertidal rocks, coralline stones
Occurrence	:	Lower mid littoral zone
Season	:	November–April
Abundance	:	Dominant
Description	:	Plants dark brown in colour, 20 – 45 cm tall with terete leafless branched rhizomatous portion and erect axes; erect axes deciduous with alternate spiral filiform branches and branchlets (Figs. a, b); vesicles elliptical innate in short branchlets; usually single rarely catnate, receptacles terete branched verrucose, linear born on third or fourth order of branching (Fig. c).

Fucales, Sargassacea

Distribution
Okha: N 22° 28.45'; E 69° 04.13'
Porbandar: N 21° 38.24'; E 69° 35.81'
Veraval: N 20° 54.87'; E 70° 20.83'

Cystoseira trinodis **(Forsskål) C. Agardh**

Thallus form and type	:	Remiform and bushy
Substratum	:	Intertidal rocks, coralline stones
Occurrence	:	Lower mid littoral zone
Season	:	November–April
Abundance	:	Dense
Description	:	Plants dark brown in colour, up to 30 cm tall or more with cylindrical muricate stems irregularly branched, branches attenuate from the base to apex (Figs. a, b); leaves present in the lower parts of the axes, mostly alternate distichous, sessile, linear, 3 – 4 cm long, 3 – 4 mm broad with mid rib, apices rounded, margins entire, vesicles innate, 1 – 3 in small branches with short intervening spaces, moniliform sub-globose, receptacles cylindrical, 3 – 4 mm long, solitary.

Fucales, Sargassaceae

Distribution
Okha: N 22° 28.43'; E 69° 04.10'
Porbandar: N 21° 38.24'; E 69° 35.81'
Shivrajpur: N 22° 19.87'; E 68° 56.95'
Veraval: N 20° 54.87'; E 70° 20.83'
Vervala: N 22° 16.87'; E 69° 56.75'

Hormophysa cuneiformis **(J. Gmelin) P. Silva**

Thallus form and type	:	Bushy with triangular shaped branches
Substratum	:	Intertidal rocks
Occurrence	:	Lower mid littoral zone
Season	:	November–April
Abundance	:	Scanty
Description	:	Plants dark brown in colour, 30 cm in height or more, sparsely branched, pseudo-dichotomously branched, branches articulated, triangular articulation narrow at the base, 3 – 5 mm broad, margins dentate, vesicles ellipsoid or oblong, embedded in the middle of swollen wings, solitary (Figs. a, b).

Fucales, Sargassaceae

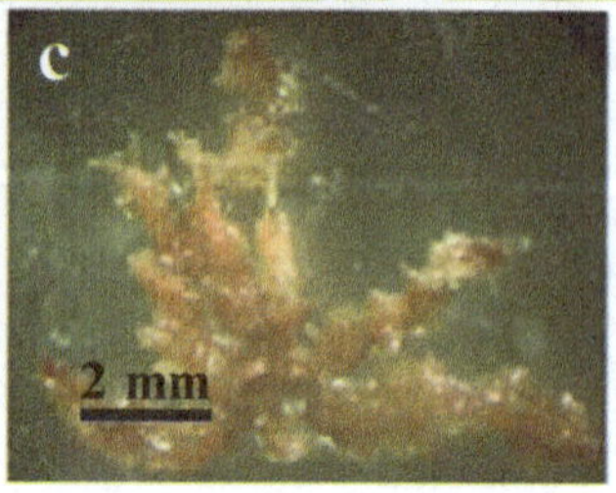

Distribution
Dwarka: N 22° 14.61'; E 68° 57.28'
Okha: N 22° 28.50'; E 69° 04.31'
Porbandar: N 21° 38.24'; E 69° 35.81'
Shivrajpur: N 22° 19.87'; E 68° 56.95'
Veraval: N 20° 54.87'; E 70° 20.83'

Sargassum cinctum **J. Agardh**

Thallus form and type	:	Foliose and bushy
Substratum	:	Intertidal rocks
Occurrence	:	Infra littoral fringe and sub littoral zone, tide pools
Season	:	November–April
Abundance	:	Dominant
Description	:	Plants dark brown in colour, up to 50 cm long, arising from disc like holdfasts; primary branches terete but compressed near the nodes; leaves linear-lanceolate or ovate, 0.3 – 2.5 cm long, 3 – 7 mm broad, serrate with obtuse apices and asymmetrical bases (Figs. a, b), costate, vesicles oval with flat winged petioles and long leaf like apicules, receptacles 5 – 10 mm long in racemose branches, serrate, spiny, ancipate-triquatrous and twisted (Fig. c).
Use	:	The species of *Sargassum* are harvested from wild and used for alginate production in India. Another note worthy product from *Sargassum* in India is liquid seaweed fertilizer.

Fucales, Sargassaceae

Distribution
Dwarka: N 22° 14.61'; E 68° 57.28'
Okha: N 22° 28.50'; E 69° 04.32'
Porbandar: N 21° 38.24'; E 69° 35.81'
Shivrajpur: N 22° 19.87'; E 68° 56.95'
Veraval: N 20° 54.87'; E 70° 20.83'

Sargassum cinereum **J. Agardh**

Thallus form and type	:	Foliose and bushy
Substratum	:	Intertidal rocks
Occurrence	:	Infra littoral fringe and sub littoral zone, tide pools
Season	:	November–April
Abundance	:	Dominant
Description	:	Plants dark brown in colour, up to 30 cm tall, leaves at the basal parts 1 – 2 cm long, 4 – 5 mm wide with rounded apices, cuneate bases, wavy, with entire margins or with coarse marginal teeth, upper leaves cuneate – lanceolate, 1 – 1.5 cm long, 2 – 3 mm broad costate, dentate, partly curved (Figs. a, b); vesicles 4 mm long, 3 – 4 mm broad, receptacles single or 2 – 3 together in a leafy and vesiculate raceme (Fig. c).
Use	:	The species of *Sargassum* are used for alginate and liquid seaweed fertilizer production in India

Fucales, Sargassaceae

Distribution
Dwarka: N 22° 14.61'; E 68° 57.28'
Okha: N 22° 28.54'; E 69° 04.59'
Porbandar: N 21° 38.24'; E 69° 35.81'
Shivrajpur: N 22° 19.87'; E 68° 56.95'
Veraval: N 20° 54.87'; E 70° 20.83'

Sargassum johnstonii **Setchell & Gardner**

Thallus form and type	:	Foliose and bushy
Substratum	:	Intertidal rocks
Occurrence	:	Infra littoral fringe and sub littoral zone, tide pools
Season	:	November–April
Abundance	:	Dominant
Description	:	Plants dark brown in colour with conical holdfast and spirally arranged terete, smooth, 4 – 8 cm long primary branches, primary branches covered with secondary branches of 2 – 12 cm long, secondary branches crowded with linear leaves; leaves thick, 1 – 2.5 cm long, 1 – 2.5 mm broad with entire to distantly denticulate margins truncate to obtuse apices (Figs. a, b); vesicles smooth born singly on short petioles or receptacular stalks, apiculate, apices cylindrical or leaf-like, male and female receptacles 2 – 5 mm long simple spindle shaped; tuberculate, often bearing leaves at the tips.

Fucales, Sargassaceae

Distribution
Dwarka: N 22° 14.61'; E 68° 57.28'
Okha: N 22° 28.57'; E 69° 04.45'
Porbandar: N 21° 38.24'; E 69° 35.81'
Shivrajpur: N 22° 19.87'; E 68° 56.95'
Veraval: N 20° 54.87'; E 70° 20.83'

***Sargassum linearifolium* (Turner) C. Agardh**

Thallus form and type	:	Foliose and bushy
Substratum	:	Intertidal rocks
Occurrence	:	Infra littoral fringe and sub littoral zone, tide pools
Season	:	November–April
Abundance	:	Abundant
Description	:	Plants dark brown in colour, 60 – 120 cm tall with terete or angular main axis. Main axis and branches spirally arranged, sparsely muricate with leaves; leaves simple, 2 – 5 cm long, 2.5 mm broad, linear, costate with acute apices and short petioles (Figs. a, b); margins sinuate, dentate sometimes with a median wing at the base or with a few teeth on the petiole; receptacles and vesicles present in the axils of the basal leaves of primary branches; receptacles stalked racemose, branches subulate or verucose, vesicles present in the fruiting branches, replacing leaves, spherical to ovate muticose, rarely apiculate, with long leaf like stalk.

Fucales, Sargassaceae

Distribution
Dwarka: N 22° 14.61'; E 68° 57.28'
Okha: N 22° 28.50'; E 69° 04.18'
Porbandar: N 21° 38.24'; E 69° 35.81'

Sargassum plagiophyllum **(Martens) J. Agardh**

Thallus form and type	:	Foliose and bushy
Substratum	:	Intertidal rocks
Occurrence	:	Infra littoral fringe and sub littoral zone, tide pools
Season	:	November–April
Abundance	:	Dense
Description	:	Plants brown in colour, 30 – 35 cm tall, attached by discoid holdfasts, richly branched with primary shoots and secondary branches (Figs. a, b); axes smooth, glabrous, cylindrical, 2 mm in thickness; leaves up to 6 cm long, 0.5 – 8.0 mm broad becoming smaller upwards; linear-oblong margins entire, sub-undulate or sub-dentate apices obtuse, midribs inconspicuous, vesicles small spherical 2 – 2.5 mm broad, receptacles racemose, stalked, verrucose, richly branched, cylindrical and spindle-shaped.

Fucales, Sargassaceae

Distribution
Okha: N 22° 29.00'; E 69° 04.40'
Porbandar: N 21° 38.24'; E 69° 35.81'
Shivrajpur: N 22° 19.87'; E 68° 56.95'

Sargassum prismaticum **Chauhan**

Thallus form and type	:	Foliose and bushy
Substratum	:	Intertidal rocks
Occurrence	:	Infra littoral fringe and sub littoral zone, tide pools
Season	:	November–April
Abundance	:	Dense
Description	:	Plants dark brown in colour, 28 – 45 cm tall, main axis and branches filiform, terete, smooth about 1 mm in diameter with spirally arranged leaves; stems at the basal parts winged, leaves linear lanceolate 5 cm long 5 – 7 mm broad with membranous with asymmetrical bases, serulate, sinuate margins, petiole short with sinuate wings, vesicles spherical, 3 – 5 mm diameter stalked, apiculate (Figs. a, b); receptacles simple, sparsely branched or 2 – 4 in a cluster with vesicles oblong, triquetrus, 5 – 11 mm, long 2 – 4 mm broad with three sinuate wings.

Fucales, Sargassaceae

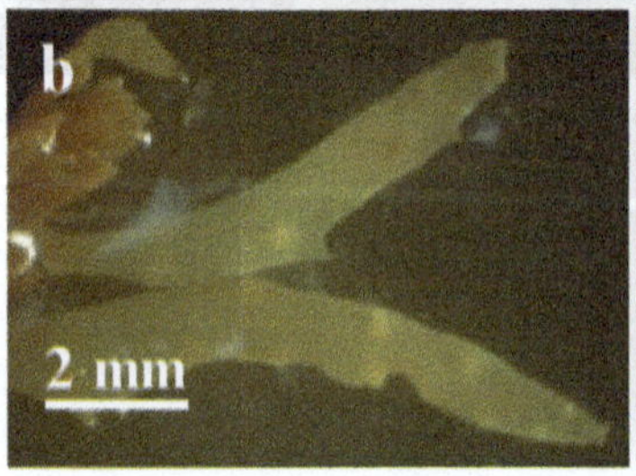

Distribution
Dwarka: N 22° 14.61'; E 68° 57.28'
Okha: N 22° 28.57'; E 69° 04.44'
Porbandar: N 21° 38.24'; E 69° 35.81'
Shivrajpur: N 22° 19.87'; E 68° 56.95'
Veraval: N 20° 54.87'; E 70° 20.83'

***Sargassum swartzii* C. Agardh**

Thallus form and type	:	Foliose and bushy
Substratum	:	Intertidal rocks
Occurrence	:	Infra littoral fringe and sub littoral zone, tide pools
Season	:	November–April
Abundance	:	Dense
Description	:	Plants dark brown in colour, 25 – 30 cm tall, attached by discoid holdfasts; thallus fleshy with flat primary and secondary branches, secondary branches alternate or distichous, leaves 3 – 7.5 cm long, 3 – 4 mm broad, sessile, cuneate at the basal parts, acuminate distantly dentate, midrib percurrent, vesicles ovate or elliptical, rarely round with flat stalks and terete apical appendages (Figs. a, b); receptacles slightly flattened, repeatedly furcated, arranged in cymose clusters, 2 – 3 apical dentate notches (Fig. c).

Fucales, Sargassaceae

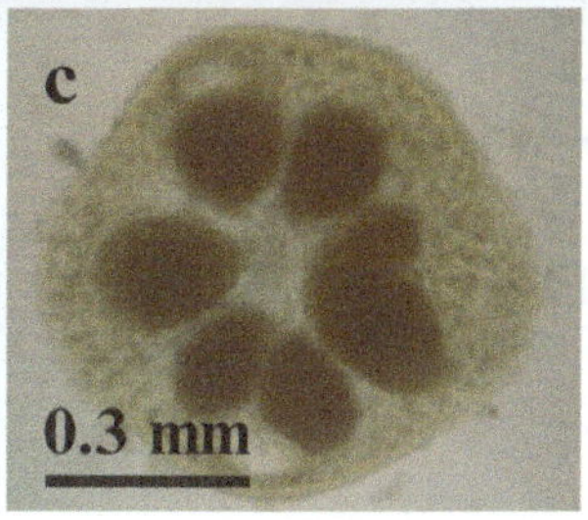

Distribution
Dwarka: N 22° 14.61'; E 68° 57.28'
Okha: N 22° 28.57'; E 69° 04.44'
Porbandar: N 21° 38.24'; E 69° 35.81'
Shivrajpur: N 22° 19.87'; E 68° 56.95'
Veraval: N 20° 54.87'; E 70° 20.83'
Vervala: N 22° 16.87'; E 69° 56.75'

Sargassum tenerrimum **J. G. Agardh**

Thallus form and type	:	Foliose and bushy
Substratum	:	Intertidal rocks
Occurrence	:	Infra littoral fringe and sub littoral zone, tide pools
Season	:	November–April
Abundance	:	Dense
Description	:	Plants yellowish brown in colour, 30 – 40 cm tall, arising form small conical discs; primary axes terete, very short, irregularly and alternately branched; primary branches cylindrical, very smooth, with branches and branchlets; leaves thin, translucent, 2 – 6 cm long, 0.5 – 1.5 cm broad, longer and broader below, smaller and narrower towards the apices, margins sinuate dentate, apices narrow rounded mid rib prominent (Figs. a, b); vesicles stalked more or less spherical; receptacles richly branched, spinous, verrucose, antheridia and oogonia developing in different receptacles (Fig. c).

Fucales, Sargassaceae

Distribution
Okha: N 22° 28.43'; E 69° 04.09'
Porbandar: N 21° 38.24'; E 69° 35.81'
Shivrajpur: N 22° 19.87'; E 68° 56.95'
Veraval: N 20° 54.87'; E 70° 20.83'

***Sargassum vulgare* C. Agardh**

Thallus form and type	:	Foliose and bushy
Substratum	:	Intertidal rocks
Occurrence	:	Infra littoral fringe and sub littoral zone, wave exposed stones
Season	:	November–April
Abundance	:	Dense
Description	:	Plants dark brown in colour, 58 cm tall with a few primary branches arising from conical holdfasts, secondary branches spirally arranged, stems terete, filiform, and sparsely muricate (Fig. a, b); leaves linear–lanceolate serrate or sub-entire in the basal parts, costate vesicles, spherical 2 – 5 mm in diameter with flat stalk, mucronate; receptacles dichotomously branched, fasciculate, cymose clusters, cylindrical, verrucose.
Use	:	The species of *Sargassum* are widely used for production of alginates and liquid seaweed fertilizer in India

Fucales, Sargassaceae

Distribution
Okha: N 22° 28.48'; E 69° 04.27'
Porbandar: N 21° 38.24'; E 69° 35.81'

Turbinaria ornata **(Turner) J. Agardh**

Thallus form and type	:	Ramified with obconical leaves
Substratum	:	Rocks
Occurrence	:	Infra littoral fringe and sub littoral zone, tide pools
Season	:	November–April
Abundance	:	Moderate
Description	:	Plants dark brown in colour, up to 50 cm tall, bushy, axes arising from dichotomously branched holdfast; main axes erect and cylindrical and irregularly branched; leaves closely arranged, turbinate to obconical, coarse, 0.5 – 1.5 cm long 10 – 15 mm broad at the distal ends (Figs. a, b); distal ends of the leaves triangular, sub-concave with double row of spines on the surface with terete stalks; vesicles immerged in the leaves; receptacles racemose, arising on the stalks of the upper leaves.
Use	:	This seaweed is used along with *Sargassum* for alginate production in India.

Rhodophyta

Bangiales, Bangiaceae

Distribution
Okha: N 22° 39.52'; E 69° 04.50'

***Porphyra* sp**

Thallus form and type	:	Leafy and blade type
Substratum	:	Intertidal rocks
Occurrence	:	Upper mid littoral zone, occasionally on splash zone
Season	:	July–September
Abundance	:	Scanty
Description	:	Plants up to 3 cm tall, 1 – 2 cm broad, foliose reddish pink in colour (Figs. a, b); monostromatic, membranous, grows as scattered tufts on rocks, attached with discoid holdfast, cells round or rectangle in shape in surface view, stellate plastid, margins with distinct spines (Fig. c)
Note	:	*Porphyra*, popularly known as nori or laver, is delicacy in Japan. *Porphyra* is rich in proteins and free amino acids besides other essential nutrients of human importance. *Porphyra* farming and processing is a big socio-economic industry in Japan, China and Korea. The annual production of *Porphyra* is about 1.5 million tons (dry wt) worldwide.

B. Jha et al., *Developments in Applied Phycology*,
DOI 10.1007/978-90-481-2488-6_3, © Springer Science + Business Media B.V. 2009

Nemaliales, Scinaiaceae

Distribution
Dwarka: N 22° 14.61'; E 68° 57.28'
Okha: N 22° 28.37'; E 69° 04.53'
Porbandar: N 21° 37.92'; E 69° 36.48'
Veraval: N 20° 54.87'; E 70° 20.83'

Scinaia carnosa **(Kützing) J. Agardh**

Thallus form and type	:	Bushy, cylindrical and succulent
Substratum	:	Intertidal rocks, rocks covered with coarse sand
Occurrence	:	Lower mid littoral zone and infra littoral fringe
Season	:	November–April
Abundance	:	Scanty
Description	:	Plants dark pinkish red in colour, up to 17 cm tall, opaque, dichotomously branched many times, slender; thallus cylindrical, deeply constricted at irregular intervals with narrow angles (Figs. a, b), central axis not visible when dry, multiaxial with central axial filaments and peripheral utricles; utricles flat topped, closely packed in a palisade fashion; cystocarps minute, scattered and not clearly visible to the naked eye.

Nemaliales, Scinaiceae

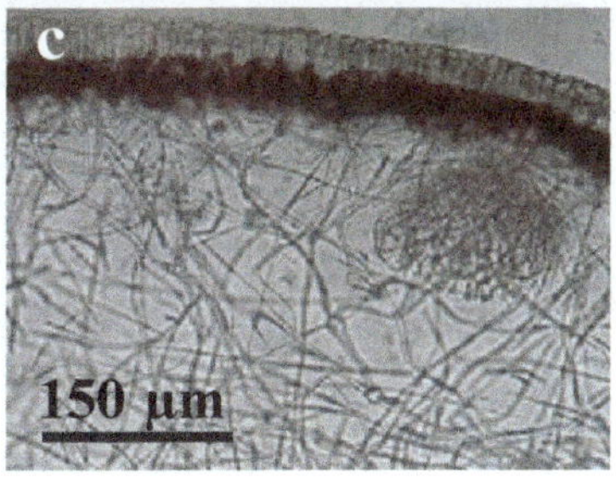

Distribution
Okha: N 22° 28.38'; E 69° 04.32' 5

Scinaia complanata **(Collins) Cotton**

Thallus form and type	:	Bushy and flattened
Substratum	:	Rocks
Occurrence	:	Lower mid littoral zone
Season	:	November–April
Abundance	:	Rare
Description	:	Plants dark red in colour, erect, up to 12 cm tall, attached by small discs; fronds flattened or complanate, dichotomously branched with round or acute apices, 3 – 5 mm broad, constrictions absent (Figs. a, b); thallus multiaxial; utricles interspersed with a few coloured cells, cystocarps scattered, immersed in the thallus (Fig. c), globular to pyriform.

Nemaliales, Scinaiaceae

a

b
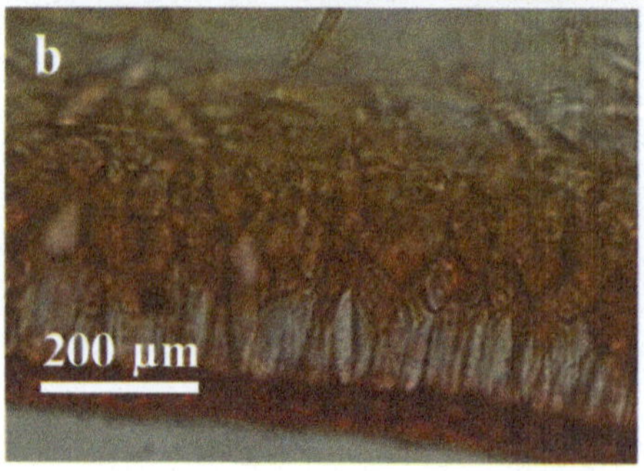

Distribution
Okha: N 22° 28.42'; E 69° 04.51'

Scinaia fascularis **(Børgesen) Huisman**

Thallus form and type	:	Bushy, succulent and cartilaginous
Substratum	:	Intertidal rocks, calcareous substrata
Occurrence	:	Lower mid littoral zone and infra littoral fringe
Season	:	July–November
Abundance	:	Scanty
Description	:	Plants bright to dark red in colour, 14 – 15 cm in height, erect, cartilaginous, attached by small discs, dichotomously branched; axes cylindrical, un-constricted, about 2 mm thick, uniform in thickness, axial strand not visible with the naked eye (Fig. a); thallus composed of filamentous central medulla and outer cortex, cortical portion consisting of pyriform and ellipsoidal utricles, intermixed with rows of pigmented cells (Fig. b).

Nemaliales, Scinaiaceae

Distribution
Dwarka: N 22° 14.61'; E 68° 57.28'
Okha: N 22° 28.50'; E 69° 04.55'
Porbandar: N 21° 37.92'; E 69° 36.48'
Veraval: N 20° 54.87'; E 70° 20.83'

***Scinaia hatei* Børgesen**

Thallus form and type	:	Bushy, cylindrical and constricted
Substratum	:	Intertidal rocks, corallines tones
Occurrence	:	Lower mid littoral zone, tide pools
Season	:	November–April
Abundance	:	Scanty
Description	:	Plants rose red in colour, cylindrical, up to 9 cm tall, erect, attached by discoid holdfast, dichotomously branched with constrictions at intervals; segments 2 – 5 mm in diameter, elongated, cylindrical of variable size, tapered below and rounded at the top (Figs. a, b); utricles flattened on the outer surface, pentagonal or hexagonal in surface view with narrow pigmented cells present between utricles in the epidermal layer.

Nemaliales, Scinaiaceae

Distribution
Dwarka: N 22° 14.61'; E 68° 57.28'
Okha: N 22° 28.48'; E 69° 04.55'
Porbandar: N 21° 37.92'; E 69° 36.48'
Veraval: N 20° 54.87'; E 70° 20.83'

Scinaia monoliformis **J. Agardh**

Thallus form and type	:	Bushy and beaded succulent fronds
Substratum	:	Intertidal rocks, coralline stones
Occurrence	:	Lower mid littoral zone, tide pools
Season	:	November–April
Abundance	:	Dominant
Description	:	Plants pinkish red in colour, up to 17 cm in height, attached by basal discs, dichotomously branched (Fig. a); branches regularly deeply constricted; segments of various sizes, elongate cylindrical in some portions, shorter, oblong oval in others, attenuated at the base and rounded above (Fig. b); utricles large, polygonal in surface view, narrow coloured present between the utricles.

Nemaliales, Galaxauraceae

Distribution
Dwarka: N 22° 14.61'; E 68° 57.28'
Okha: N 22° 28.43'; E 69° 04.10'
Porbandar: N 21° 37.92'; E 69° 36.48'
Shivrajpur: N 22° 19.87'; E 68° 56.95'
Veraval: N 20° 54.87'; E 70° 20.83'
Vervala: N 22° 16.87'; E 69° 56.75'

***Tricleocarpa fragilis* (Linnaeus) Huisman & Townsend**

Thallus form and type	:	Bushy and calcified
Substratum	:	Intertidal rocks, calcareous substrata
Occurrence	:	Mid littoral zone, tide pools
Season	:	November–April
Abundance	:	Moderate
Description	:	Plants olive red in colour, 5 – 6 cm tall, dimorphic, male plants much smaller than female plants; branching regularly dichotomous with articulations in basal positions; branches cylindrical uniform with smooth surface, 1 – 2 mm in diameter; growing points situated in a deep narrow depressions at apices (Figs. a, b); medullary filaments loosely interwoven; calcification strong, confined to the cortex in the younger parts, extending to the peripheral portion of the medulla in the older portions.

Nemaliales, Liagoraceae

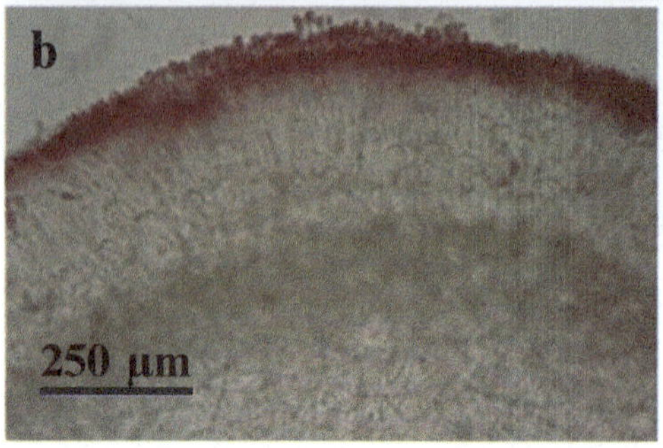

Distribution
Adri: N 20° 57.58'; E 70° 16.76'

Dermonema virens **(J. Agardh) Pedrohche & Avila Ortiz**

Thallus form and type	:	Bushy, cylindrical and succulent
Substratum	:	Intertidal rocks, corallines tones
Occurrence	:	Mid – Upper littoral zone, tidal pools
Season	:	November–April
Abundance	:	Scanty
Description	:	Plants dark brown to red in colour, un-calcified, somewhat gelatinous, 6 – 10 cm in height with several axes arising from a basal disc; erect axes dichotomously and closely branched gradually tapering from the base to the tip(s) (Fig. a); thallus multiaxial with filamentous medulla and cortex (Fig. b).
Use	:	This alga used in food

Nemaliales, Ligoraceae

Distribution
Adri: N 20° 57.58'; E 70° 16.76'
Dwarka: N 22° 14.61'; E 68° 57.28'
Okha: N 22° 28.42'; E 69° 04.11'
Porbandar: N 21° 37.92'; E 69° 36.48'
Veraval: N 20° 54.87'; E 70° 20.83'

Helminthocladia clavadosii **(Lamouroux ex Duby) Satchell f.** ***indica*** **Desikachary**

Thallus form and type	:	Bushy and mucilaginous
Substratum	:	Intertidal rocks, calcareous substrata
Occurrence	:	Lower mid littoral zone, tidal streams and pools
Season	:	November–April
Abundance	:	Scanty
Description	:	Plants reddish purple in colour, 30 – 50 cm tall, attached by small basal discs, simple or branched (Figs. a, b); thallus multi-axial with radially arranged cortical filaments, assimilatory filaments 4 – 5 times divided, terminal cells of the cortical filaments large pyriform (Fig. c); rhizoids up to cortical filaments.

Nemaliales, Ligoraceae

Distribution
Dwarka: N 22° 14.61'; E 68° 57.28'
Okha: N 22° 28.42'; E 69° 04.11'
Porbandar: N 21° 37.92'; E 69° 36.48'
Shivrajpur: N 22° 19.87'; E 68° 56.95'

Liagora ceranoides **Lamouroux**

Thallus form and type	:	Bushy and mucilaginous
Substratum	:	Intertidal rocks and dead corals
Occurrence	:	Mid littoral zone, tide pools and streams
Season	:	November–April
Abundance	:	Moderate
Description	:	Plants whitish to reddish purple in colour, up to 8 cm tall, bushy; thallus calcified, slightly compressed or terete, repeatedly dichotomously branched with few proliferations; calcification uniform, apices of branches somewhat divaricate (Figs. a, b); assimilatory filaments up to eight times forked; cells moniliform to subcylindrical.

Nemaliales, Ligoraceae

Distribution
Shivrajpur: N 22° 19.87'; E 68° 56.95'

Liagora viscida **(Forsskål) C. Agardh**

Thallus form and type	:	Bushy and corymbose
Substratum	:	Intertidal rocks, dead corals, coralline stones
Occurrence	:	Mid littoral zone, tidal pools and streams
Season	:	November–April
Abundance	:	Scanty
Description	:	Plants whitish to reddish purple in colour, tips rose red, 2 – 4 cm tall, attached by small basal discs (Fig. a); fronds small, viscid, calcified, closely dichotomously branched; angles between the branches somewhat divergent.

Ahnfeltiales, Ahnfeltiaceae

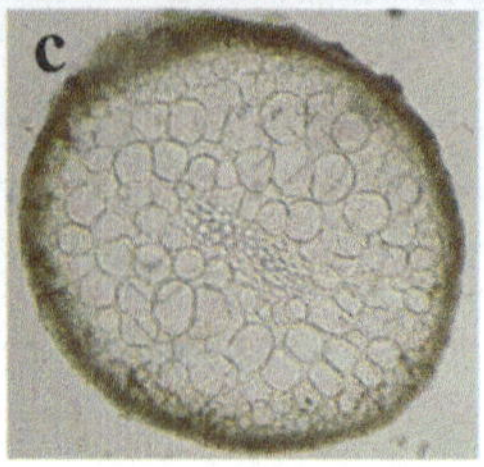

Distribution
Kotda: N 20° 42.38'; E 70° 52.17'

***Ahnfeltia plicata* (Hudson) Fries**

Thallus form and type	:	Bushy, wiry and corymbose
Substratum	:	Intertidal rocks
Occurrence	:	Lower mid littoral zone, intertidal rock pools and often found buried in sand
Season	:	December–April
Abundance	:	Moderate
Description	:	Plants forms tangled dense tufts, blackish red in colour, solitary, upper most branches slightly green in colour, about 15 cm tall, fronds erect, stiff and wiry with irregular or dichotomously branching (Fig. a, b). The holdfast is disc-like or encrusting, 0.5 – 1.5 cm in diameter.
Use	:	Ahnfeltia is one of the world's principal commercial agarophyte. Russia annually harvests about 5,000 t (dry wt) from wild stocks and uses for production of high quality low sulphate agar.

Gelidiales, Gelidiaceae

Distribution
Dwarka: N 22° 14.61'; E 68° 57.28'
Kotda: N 20° 42.38'; E 70° 52.17'
Okha: N 22° 28.42'; E 69° 04.19'
Porbandar: N 21° 37.33'; E 69° 36.48'
Shivrajpur: N 22° 19.87'; E 68° 56.95'
Veraval: N 20° 54.87'; E 70° 20.83'

Gelidium micropterum **Kützing**

Thallus form and type	:	Flattened thallus tufts
Substratum	:	Calcareous stones
Occurrence	:	Lower mid littoral zone and infra littoral fringe
Season	:	November–April
Abundance	:	Moderate
Description	:	Plants dark red in colour, up to 5 cm tall, growing as tufts or patches, cartilaginous, flattened 0.5 – 2.0 mm broad with blunt apices and sunken growing points; branching bi- or tri-pinnate or irregular sometimes (Figs. a, b); tetrasporangia confined to short swollen fertile tips with rounded ends, scattered all over the surface of the fertile portion; cystocarps swollen, sub-spherical giving fertile tips of the branchlets a blunt appearance.
Use	:	*Gelidium* is an important agarophyte in India. The poor biomass and stunted growth make it unattractive.

Gelidiales, Gelidiaceae

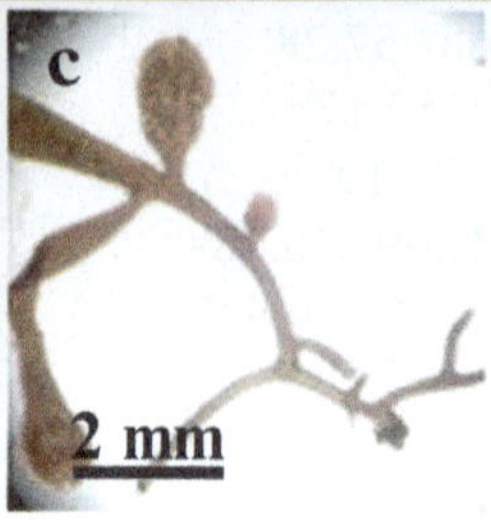

Distribution
Dwarka: N 22° 14.61'; E 68° 57.28'
Okha: N 22° 28.41'; E 69° 04.16'
Porbandar: N 21° 37.92'; E 69° 36.48'
Shivrajpur: N 22° 19.87'; E 68° 56.95'
Veraval: N 20° 54.87'; E 70° 20.83'

Gelidium pusillum **(Stackhouse) Le Jolis**

Thallus form and type	:	Flattened tufts
Substratum	:	Intertidal rocks, corallines tones
Occurrence	:	Upper mid littoral zone, sheltered
Season	:	November–April
Abundance	:	Moderate
Description	:	Plants dark red in colour, 1 – 1.5 cm tall, small, cartilaginous, forming short tufts or matted into cushions (Figs. a, b); creeping stolons giving rise to erect blades attached at frequent intervals by rhizoids or discoid heptera; fronds sub-cylindrical below and compressed above, simple or rarely irregularly to sub-pinnately branched (Fig. c); carpogonial sori in the terminal portions of the branchlets.

Gelidiales, Gelidiellaceae

Distribution
Adri: N 20° 57.58'; E 70° 16.76'
Dwarka: N 22° 14.61'; E 68° 57.28'
Kotda: N 20° 42.38'; E 70° 52.17'
Okha: N 22° 28.33'; E 69° 03.51'
Porbandar: N 21° 37.33'; E 69° 36.48'
Veraval: N 20° 54.87'; E 70° 20.83'

Gelidiella acerosa **(Forsskål) J. Feldmann**

Thallus form and type	:	Remiform and cartilaginous
Substratum	:	Rocks, coralline stones and calcareous substrata
Occurrence	:	Lower mid littoral zone, tidal flat pools
Season	:	November–April
Abundance	:	Moderate
Description	:	Plants dull purplish to yellowish in colour, 5 – 7 cm tall, cylindrical, rigid and cartilaginous in texture, loosely matted with decumbent creeping portions and erect fronds (Fig. a); thallus irregularly branched, some times branches regularly pinnate or bi-pinnate (Fig. b); in young portions long unicellular hairs developing near tips of erect branches; stichidia are is forming in upper apices of pinnules, stichidia irregularly clumped without any definite arrangement. Attempts are underway to cultivate this alga in the sea on large scale.
Use	:	*Gelidiella* forms an important source of raw material for production of bacteriological grade agar in India. The entire industrial requirement is met from the wild harvest which is estimated at 300 t dry weight per annum. Jelly prepared from agar is consumed as food.

Gracilariales, Gracilariaceae

Distribution
Adri: N 20° 57.58'; E 70° 16.75'
Dwarka: N 22° 14.61'; E 68° 57.28'
Kotda: N 20° 42.38'; E 70° 52.17'
Okha: N 22° 28.44'; E 69° 03.58'
Porbandar: N 21° 37.92'; E 69° 36.48'
Shivrajpur: N 22° 19.87'; E 68° 56.95'
Veraval: N 20° 54.87'; E 70° 20.83'

***Gracilaria corticata* (J. Agardh) J. Agardh**

Thallus form and type	:	Bushy and cartilaginous
Substratum	:	Intertidal rocks, calcareous substrata
Occurrence	:	Lower mid littoral zone and infra littoral fringe
Season	:	November–April
Abundance	:	Dominant
Description	:	Plants dark red to yellowish red in colour, bushy, 9 – 17 cm in height, rigid, cartilaginous, flattened, dichotomously branched with narrow segments; segments usually 2 – 4 mm wide, tips of segments acute, sometimes with proliferations (Figs. a, b); thallus composed of 1 – 2 layered cortex and medulla of large cells at the centre.
Use	:	Species of *Gracilaria* are harvested from wild stocks and used for preparing food grade agar in India. There are some efforts being made to cultivate on large scale.
18S rRNA gene sequence	:	NCBI Gen Bank Accession No. DQ 409340
*cox*2 - *cox*3 spacer	:	NCBI Gen Bank Accession No. EU 937758
RUBISCO spacer	:	NCBI Gen Bank Accession No. EU 937767
ITS1	:	NCBI Gen Bank Accession No. EU 937776

Gracilariales, Gracilariaceae

Distribution
Adri: N 20° 57.58'; E 70° 16.76'
Dwarka: N 22° 14.61'; E 68° 57.28'
Kotda: N 20° 42.38'; E 70° 52.17'
Navabandar: N 22° 45.33'; E 69° 47.22'
Okha: N 22° 28.42'; E 69° 03.58'
Porbandar: N 21° 37.92'; E 69° 36.48'
Rajapara: N 20° 47.46'; E 71° 12.16'

***Gracilaria corticata* (J. Agardh) J. Agardh var. *cylindrica* Umamaheswara Rao**

Thallus form and type	:	Bushy and cartilaginous
Substratum	:	Intertidal boulders, coralline substrata
Occurrence	:	Lower mid littoral zone and infra littoral fringe
Season	:	November–April
Abundance	:	Moderate
Description	:	Plants dark red to yellowish red in colour, bushy, 10 cm tall or more, rigid, cartilaginous, often with variegated spots on fronds; fronds dichotomously, alternately and irregularly branched, compressed at the lower parts, sub-terete to cylindrical and closely branched above with pointed or spinous apices (Figs. a, b).
18S rRNA gene sequence	:	NCBI Gen Bank Accession No. DQ 409339
*cox*2-*cox*3 spacer	:	NCBI Gen Bank Accession No. EU 937759
RUBISCO spacer	:	NCBI Gen Bank Accession No. EU 937768
ITS1	:	NCBI Gen Bank Accession No. EU 937777

Gracilariales, Gracilariaceae

Distribution
Vervala: N 22° 16.87'; E 69° 56.75'

Gracilaria debilis **(Forsskål) Børgesen**

Thallus form and type	:	Bushy and succulent
Substratum	:	Intertidal rocks and calcareous substrata
Occurrence	:	Upper mid littoral zone, tide pools
Season	:	December–April
Abundance	:	Scanty
Description	:	Plants dark green to blackish red in colour, up to 10 cm in height, cylindrical up to 3 mm thickness, di- or sub-dichotomously branched at the upper ends of the axes; fronds sub-fastigiate and ramified at the distal ends (Figs. a, b); cortex 1 – 2 layered, outer layer of elongated cells and inner layer of radially compressed cells; medulla with large central cells (Fig. c); cystocarps dome shaped, scattered over the mature parts of the thallus.
18S rRNA gene sequence	:	NCBI Gen Bank Accession No. DQ 937745
*cox*2-*cox*3 spacer	:	NCBI Gen Bank Accession No. EU 937764
RUBISCO spacer	:	NCBI Gen Bank Accession No. EU 937773
ITS1	:	NCBI Gen Bank Accession No. EU 937782

Gracilariales, Gracilariaceae

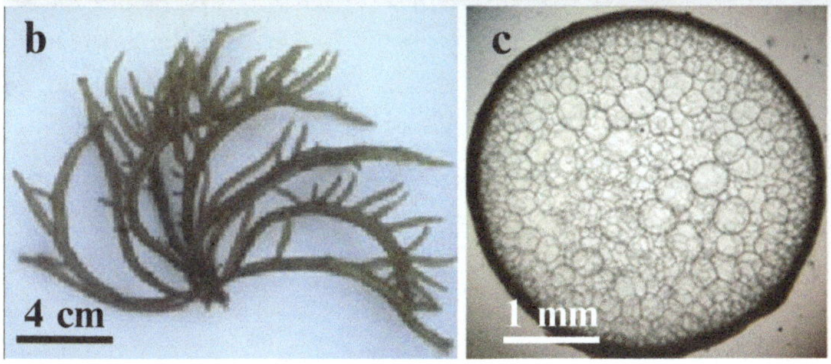

Distribution
Adri: N 20° 57.58'; E 70° 16.76'
Okha: N 22° 28.38'; E 69° 03.56'
Veraval: N 20° 54.87'; E 70° 20.83'

Gracilaria dura **(C. Agardh) C. Agardh**

Thallus form and type	:	Bushy and succulent
Substratum	:	Intertidal rocks
Occurrence	:	Tide pools, intertidal streams
Season	:	November–May
Abundance	:	Moderate at Viraval and scantily available in other coasts
Description	:	Plants dark green to yellowish green in colour, 10 – 18 cm tall, bushy, cylindrical, irregularly branched, attached to rocks by discoid holdfast; branches are often covered with short branchlets, which are constricted at the base, pointed or forked at the apices (Figs. a, b); thallus composed of single layer of cortical cells and central medulla of large ovoid cells (Fig. c). Efforts are under way to cultivate this alga in field using raft method.
Use	:	Source of high gel strength agar and agarose (U.S. Patent No. 2005/0267296 A1; WO 2005/118830).
18S rRNA gene sequence	:	NCBI Gen Bank Accession No. DQ 399795
*cox*2-*cox*3 spacer	:	NCBI Gen Bank Accession No. EU 937757
RUBISCO spacer	:	NCBI Gen Bank Accession No. EU 937766
ITS1	:	NCBI Gen Bank Accession No. EU 937775

Gracilariales, Gracilariaceae

Distribution
Veraval: N 20° 54.87'; E 70° 20.83'

Gracilaria fergusonii **J. Agardh**

Thallus form and type :	Ramiform, bushy succulent
Substratum :	Intertidal rocks
Occurrence :	Mid littoral zone, rock pools
Season :	November–April
Abundance :	Scanty
Description :	Plants about 10 cm tall, unbranched up to about one half of its height and then repeatedly dichotomously branched to form a flabellate or fastigiated crown (Figs. a, b), ultimate branches slightly compressed, apices rounded.
18S rRNA gene sequence :	NCBI Gen Bank Accession No. EU 937740
*cox*2-*cox*3 spacer :	NCBI Gen Bank Accession No. EU 937760
RUBISCO spacer :	NCBI Gen Bank Accession No. EU 937769
ITS1 :	NCBI Gen Bank Accession No. EU 937778

Gracilariales, Gracilariaceae

Distribution
Dwarka: N 22° 14.61'; E 68° 57.28'
Okha: N 22° 28.4'; E 69° 03.58'
Porbandar: N 21° 37.92'; E 69° 36.48'
Veraval: N 20° 54.87'; E 70° 20.83'

Gracilaria foliifera **(Forsskål) Børgesen**

Thallus form and type	:	Bushy and leathery
Substratum	:	Intertidal rocks, calcareous substrata
Occurrence	:	Lower mid littoral zone, tide pools
Season	:	November – April
Abundance	:	Scanty
Description	:	Plants brownish to yellowish red in colour, up to 16 cm in height, flat or compressed dichotomously or sub-dichotomously branched, attached by a discoid holdfast; frond membranous, proliferous, and branches tapering towards the apices, laciniate with acuminate tips, proliferations often marginal (Figs. a, b).
18S rRNA gene sequence	:	NCBI Gen Bank Accession No. EU 937741.
*cox*2-*cox*3 spacer	:	NCBI Gen Bank Accession No. EU 937761.
RUBISCO spacer	:	NCBI Gen Bank Accession No. EU 937770.
ITS1	:	NCBI Gen Bank Accession No. EU 937779.

Gracilariales, Gracilariaceae

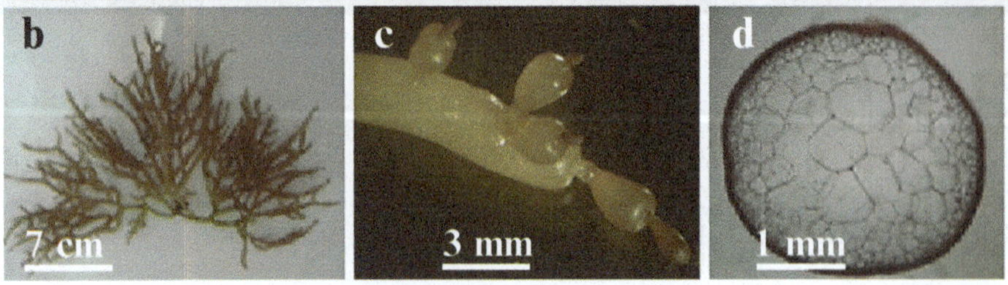

Distribution
Shivrajpur: N 22° 19.87';
E 68° 56.95'

Gracilaria salicornia **(C. Agardh) Dawson**

Thallus form and type	:	Bushy and succulent
Substratum	:	Intertidal rocks
Occurrence	:	Lower mid littoral zone, rock pools
Season	:	November–April
Abundance	:	Moderate
Description	:	Plant brownish to yellowish red in colour, up to 16cm in height, attached by small discs; thallus bushy with irregularly branched, cylindrical axes; lower branches cylindrical, not attenuated at the base (Figs. a, b): upper branches attenuated below, elongate clavate, swollen at the apex and showing apical depressions, one or two branchlets arising from the depressions (Fig. c); cortex 1 – 2 layered, medulla with large central cells (Fig. d); cystocarps scattered all over the thallus, rostrate.
18S rRNA gene sequence	:	NCBI Gen Bank Accession No. EU 937743
*cox*2-*cox*3 spacer	:	NCBI Gen Bank Accession No. EU 937763
RUBISCO spacer	:	NCBI Gen Bank Accession No. EU 937772
ITS1	:	NCBI Gen Bank Accession No. EU 937781

Gracilariales, Gracilariaceae

Distribution
Kotada: N 20° 42.38'; E 70° 52.17'
Okha: N 22° 28.38'; E 69° 03.56'
Shivrajpur: N 22° 19.87'; E 68° 56.95'

Gracilaria textorii **(Suringar) De Toni**

Thallus form and type	:	Foliose and branched
Substratum	:	Intertidal rocks
Occurrence	:	Lower mid littoral zone, tide pools
Season	:	November–April
Abundance	:	Moderate
Description	:	Plants yellowish in colour, 9 – 35 cm tall, flat, foliose, membranous to thick coriaceous, ribbon-like dichotomously, sub-dichotomously or irregularly branched forming a flabellate expanse; branches 0.5 – 2.0 cm broad with rounded or somewhat attenuated apices; margins entire or proliferous (Figs. a, b); cystocarps prominent emergent, globose and scattered all over the thallus.
18S rRNA gene sequence	:	NCBI Gen Bank Accession No. EU 937746
*cox*2-*cox*3 spacer	:	NCBI Gen Bank Accession No. EU 937765
RUBISCO spacer	:	NCBI Gen Bank Accession No. EU 937774
ITS1	:	NCBI Gen Bank Accession No. EU 937783

Gracilariales, Gracilariaceae

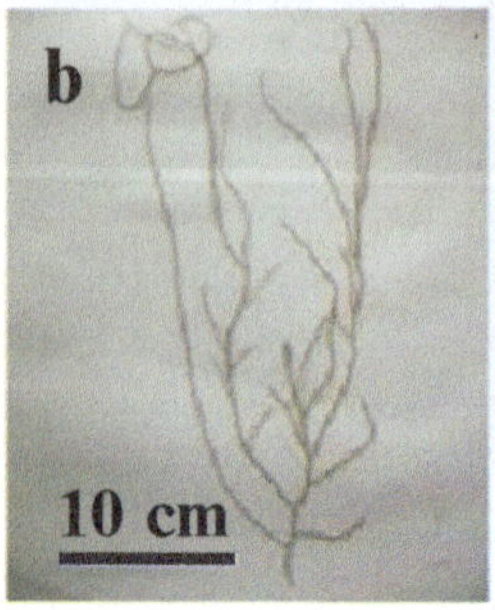

Distribution
Okha: N22° 26.38'; E 69° 03.30'
Sikka: N 22° 26.23'; E 69° 49.70'
Shivrajpur: N 22° 19.87'; E 68° 56.95'

Gracilaria verrucosa **(Hudson) Papenfuss**

Thallus form and type	:	Ramiform, cylindrical
Substratum	:	Intertidal stones, pebbles or shells
Occurrence	:	Lower mid littoral zone, tide pools
Season	:	November–April
Abundance	:	Moderate
Description	:	Plants yellowish to brownish red in colour, up to 30 cm tall, erect, terete, attached to small stones by small circular discs, branching lateral, sub-dichotomous, alternate, branch tips attenuated and ultimate branches small, with branches up to 3rd or 4th order (Figs. a, b); cystocarps sub-spherical, elevated and scattered over the thallus (Fig. c).
18S rRNA gene sequence	:	NCBI Gen Bank Accession No. EU 937742
*cox*2-*cox*3 spacer	:	NCBI Gen Bank Accession No. EU 937762
RUBISCO spacer	:	NCBI Gen Bank Accession No. EU 937771
ITS1	:	NCBI Gen Bank Accession No. EU 937780

Bonnemaisoniales, Bonnemaisoniaceae

Distribution
Veraval: N 20° 54.87'; E 70° 20.83'
Samiani Island: N 22° 29.10'; E 69° 05.03'

***Asparagopsis taxiformis* (Delile) Trevisan**

Thallus form and type	:	Fronds with brush-like tufts
Substratum	:	Bed rocks
Occurrence	:	Sub littoral zones
Season	:	February–April
Abundance	:	Scanty
Description	:	Plants dark red to purple red in colour, 15 – 20 cm tall with creeping branched rhizomatous portion and rhizoidal holdfast; erect shoots arising from the creeping rhizomes, naked below and densely covered with brush-like branches and branchlets at the upper part (Figs. a, b), bushy or plumose; branches sub-opposite pinnate and provided with delicate branchlets, ultimate branchlets filamentous with three rows of cells.
Use	:	Iodine content of this species has been reported to be 0.5% on dry weight basis. This feature has attracted some researchers in India to study the feasibility of its extraction and its resource survey

Cryptonemiales, Corynomorphaceae

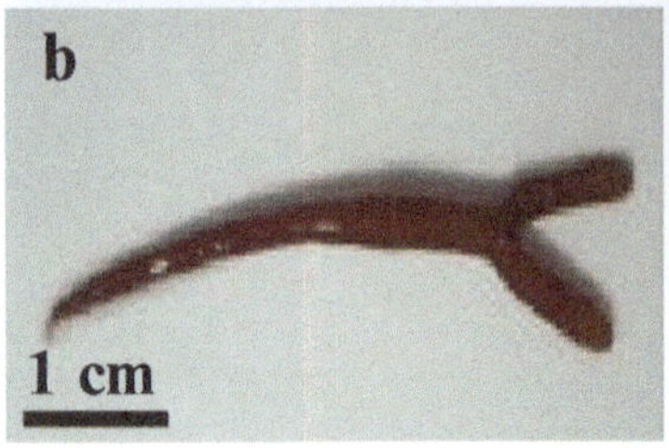

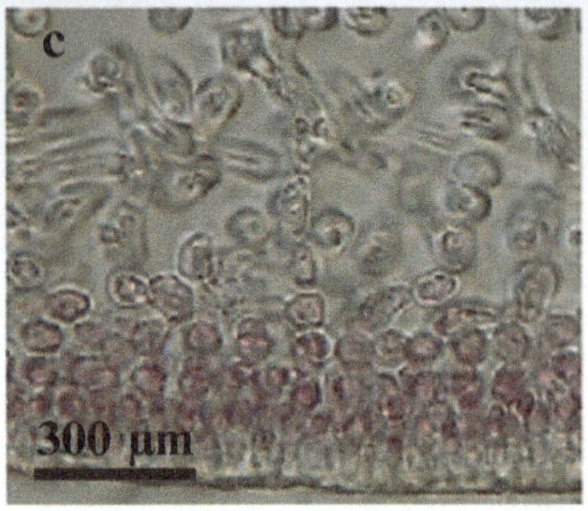

Distribution
Veraval: N 20° 54.87';
E 70° 20.83'

***Corynomorpha prismatica* (J. Agardh) J. Agardh**

Thallus form and type	:	Cylindrical and sub-cartilaginous
Substratum	:	Intertidal rocks
Occurrence	:	Supra littoral zone, mostly in tide pools
Season	:	July–November
Abundance	:	Scanty
Description	:	Plants pink to dark in red colour, 8 – 10 cm tall or more, attached by small holdfasts, erect, axes stipitate, 1 – 1.5 mm thick at the basal portion, 3 – 4 mm in diameter above, clavate with rounded tips (Figs. a, b), simple or sub-dichotomously branched; thallus composed of a filamentous medulla and a compact cortex (Fig. c).

Cryptonemiales, Peyssonneliaceae

a

Distribution
Adri: N 20° 57.58'; E 70° 16.76'

***Peyssonnelia obscura* Weber na Bosse var. *bombayensis* Børgesen**

Thallus form and type	:	Crustose
Substratum	:	Coralline stones
Occurrence	:	Lower mid littoral zone
Season	:	April
Abundance	:	Scanty
Description	:	Plants dark red in colour, crustose, orbicular, attached by unicellular rhizoids arising from the lower surface (Fig. a); margins free, thallus composed of single row of hypothelial cells and vertical rows of perithalial filaments; perithallial filaments 4 – 6 celled.

Halymeniales, Halymeniaceae

Distribution
Okha: N 22° 28.48'; E 69° 04.56'
Veraval: N 20° 54.87'; E 70° 20.83'

Cryptonemia undulata **Sonder**

Thallus form and type	:	Leafy and frondose
Substratum	:	Bed rocks
Occurrence	:	Sub littoral zone
Season	:	February–April
Abundance	:	Scanty
Description	:	Plants dark red in colour, 12 – 15 cm high, foliose, firm, elastic with basal attachment discs; fronds, erect, caespitose, stipitate, stipe rigid, expanding into oblong linear blades up to three times divided; angles between lobes rather obtuse; blades with prominent midrib, 2 – 3 cm broad with undulate and lobate margins, blunt apices; proliferations common, variable in number, arising from the midrib, obovate, oblong and enervate (Figs. a, b).

Halymeniales, Halymeniaceae

Distribution
Dwarka: N 22° 14.61'; E 68° 57.28'
Okha: N 22° 28.43'; E 69° 04.10'
Porbandar: N 21° 37.92'; E 69° 36.48'
Shivrajpur: N 22° 19.87'; E 68° 56.95'
Sutrapada: N 20° 50.18'; E 70° 28.89'
Veraval: N 20° 54.87'; E 70° 20.83'

Grateloupia filicina **(Lamouroux) C. Agardh**

Thallus form and type	:	Bushy and mucilaginous
Substratum	:	Intertidal rocks, static coarse sand beds
Occurrence	:	Upper mid littoral zone
Season	:	November–April
Abundance	:	Scanty
Description	:	Plants brownish green or blackish red in colour, 8 – 12 cm long, erect, lubricous, sub-cylindrical or compressed; branches simple in the upper portion of the thallus, monopodial or variable in the basal part; marginal branchlets sub-distichous and somewhat narrow (Figs. a, b).
Use	:	*Grateloupia* species are used as food. Jelly prepared from this alga is consumed as human food in Korea. *G. filicina* is popularly known as "Centipede algae"

Halymeniales, Halymeniaceae

Distribution
Dwarka: N 22° 14.61'; E 68° 57.28'
Okha: N 22° 28.50'; E 69° 04.55'
Porbandar: N 21° 37.92'; E 69° 36.48'
Shivrajpur: N 22° 19.87'; E 68° 56.95'
Sutrapada: N 20° 50.18'; E 70° 28.89'
Veraval: N 20° 54.87'; E 70° 20.83'

***Grateloupia indica* Børgesen**

Thallus form and type	:	Leafy and lubricous
Substratum	:	Intertidal rocks, calcareous stones
Occurrence	:	Mid littoral zone, rocks surrounding tide pools
Season	:	November–April
Abundance	:	Moderate
Description	:	Plants dark red in colour, up to 100 cm in height, foliose, tough in consistency with lubricous surface, attached to rocks by small basal discs; lobes of the frond irregular with sinuate or wavy margins (Figs. a, b); older fronds have numerous proliferations; perforated frond also proliferous; proliferations marginal, small and spathulate.

Halymeniales, Halymeniaceae

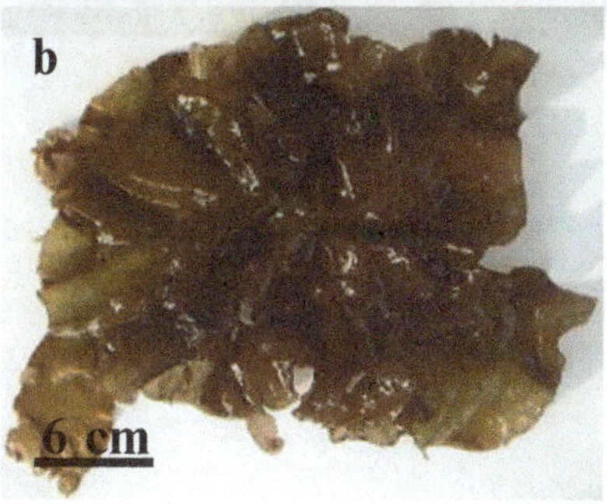

Distribution
Dwarka: N 22° 14.61'; E 68° 57.28'
Okha: N 22° 28.51'; E 69° 04.50'
Porbandar: N 21° 37.92'; E 69° 36.48'
Samiani Island: N 22° 29.30'; E 69° 05.20'
Shivrajpur: N 22° 19.87'; E 68° 56.95'
Vceraval: N 20° 54.87'; E 70° 20.83'

***Halymenia porphyraeformis* Parkinson**

Thallus form and type	:	Leafy and mucilaginous
Substratum	:	Intertidal rocks
Occurrence	:	Lower mid littoral zone and infra littoral fringe
Season	:	November–April
Abundance	:	Scanty
Description	:	Plants rose red in colour, up to 48 cm long, 40 cm wide, erect, tough and elastic in consistency, cordate base and sinuate or wavy margins; fronds sometimes lobed in the upper position (Figs. a, b); tetrasporangia ovoid to sub-spherical, cruciately divided and scattered in the cortical layers; cystocarps sub-spherical, scattered in the thallus and extending out into the medullary region.
Use	:	*Halymenia* species are used for food

Halymeniales, Halymeniaceae

Distribution
Dwarka: N 22° 14.61'; E 68° 57.28'
Okha: N 22° 28.50'; E 69° 04.46'
Porbandar: N 21° 37.92'; E 69° 36.48'
Samiani Island: N 22° 29.30'; E 69° 05.20'
Shivrajpur: N 22° 19.87'; E 68° 56.95'
Veraval: N 20° 54.87'; E 70° 20.83'

***Halymenia venusta* Børgesen**

Thallus form and type	:	Leafy and lubricous
Substratum	:	Intertidal rocks
Occurrence	:	Lower mid littoral zone and infra littoral fringe
Season	:	November–April
Abundance	:	Moderate
Description	:	Plants light rose red in colour, up to 30 cm tall, slightly lubricous, large and attached by small basal discs; fronds foliose, variable in shape, palmately divided into several lobes; margins and surface of the blades are densely proliferated; proliferations vary from many centimeters to a few millimeters in size, repeatedly divided, narrow with sharply acute tips (Figs. a, b).

Corallinales, Corallinaceae

Distribution
Dwarka: N 22° 14.61'; E 68° 57.28'
Kotda: N 20° 42.38'; E 70° 52.17'
Okha: N 22° 28.48'; E 69° 04.36'
Porbandar: N 21° 37.92'; E 69° 36.48'
Shivrajpur: N 22° 19.87'; E 68° 56.95'
Veraval: N 20° 54.87'; E 70° 20.83'

***Amphiroa anceps* (Lamark) Decaisne**

Thallus form and type	:	Calcified and articulated
Substratum	:	Corraline stones
Occurrence	:	Sub littoral zone, tide pools
Season	:	November–April
Abundance	:	Moderate
Description	:	Plants red to bright purple in colour, 5 – 10 cm tall, erect articulate with strongly calcified segments (Fig. a) or intergenucula and uncalcified genicula/joints, irregularly dichotomous, sub-trichotomous in the lower portions with flat segments; edges of the branches uneven and apices obtuse (Fig. b); branches arising from the broadened tips of the intergenicula; intergenicula or segments flattened and occasionally sub-terete in lower portion; conceptacles only on one side of the intergenicula, occasionally on both sides, slightly prominent.

Corallinales, Corallinaceae

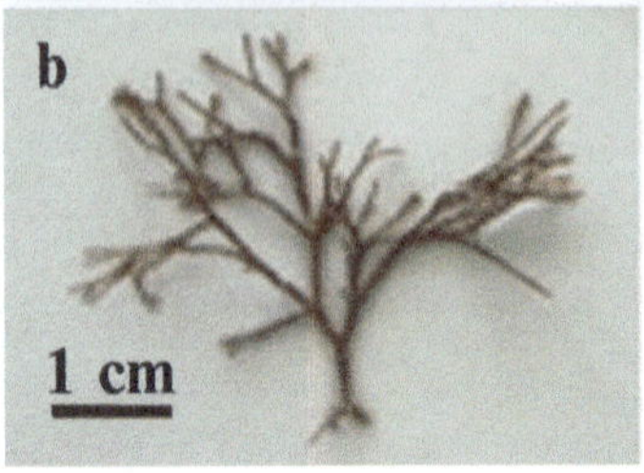

Distribution
Gopnath: N 21° 12.52'; E 72° 06.53'
Porbandar: N 21° 37.92'; E 69° 36.48'
Veraval: N 20° 54.87'; E 70° 20.83'

***Amphiroa fragilissima* (Linnaeus) Lamouroux**

Thallus form and type	:	Calcified and articulated
Substratum	:	Coralline stones
Occurrence	:	Lower mid littoral zone, tide pools
Season	:	November–April
Abundance	:	Moderate
Description	:	Plants purple red in colour, up to 3 cm tall, calcified, erect, fragile, regularly dichotomously or trichotomously branched, sometimes with adventitious branches, apices obtuse (Figs. a, b); segments or intergenicula cylindrical or slightly compressed, several times longer than broad, sometimes with pad-like swellings at the tip; conceptacles lateral, hemispherical, prominent.

Corallinales, Corallinaceae

Distribution
Dwarka: N 22° 14.61'; E 68° 57.28'
Kotda: N 20° 42.15'; E 70° 52.07'
Porbandar: N 21° 37.92'; E 69° 36.48'
Veraval: N 20° 54.87'; E 70° 20.83'

Cheilosporum spectabile **Harvey ex Grunow**

Thallus form and type	:	Bushy and calcified
Substratum	:	Intertidal rocks, coralline stones
Occurrence	:	Tide pool
Season	:	November–April
Abundance	:	Moderate
Description	:	Plants pink to red in colour, up to 6 cm tall, delicate, erect and articulated, regularly dichotomously branched; intergenicula multizonal, markedly flattened with prominent midrib and wings on both sides (Figs. a, b); conceptacles solitary or in pairs, confined to the acute tips of the wings.

Corallinales, Corallinaceae

Distribution
Dwarka: N 22° 14.61'; E 68° 57.28'
Okha: N 22° 28.50'; E 69° 04.33'
Porbandar: N 21° 37.92'; E 69° 36.48'
Shivrajpur: N 22° 19.87'; E 68° 56.95'
Veraval: N 20° 54.87'; E 70° 20.83'

***Corallina berteroi* Montagne ex Kützing**

Thallus form and type	:	Bushy and calcified
Substratum	:	Intertidal rocks, coralline stones
Occurrence	:	Tide pool
Season	:	November–April
Abundance	:	Moderate
Description	:	Plants light purple in colour, up to 3 – 4 cm tall, erect, calcified, articulated, erect fronds cylindrical to flattened, dichotomously branched arising from a small, circular, basal disc; branching regular, pinnate or dichotomous, approaching cymoid, with many transitory types, usually somewhat lax; small proliferations arise from the segments or intergenicula; ultimate intergenicula with spatulate tips (Figs. a, b); tetrasporic conceptacles borne in small terminal cymoid clusters.

Corallinales, Corallinaceae

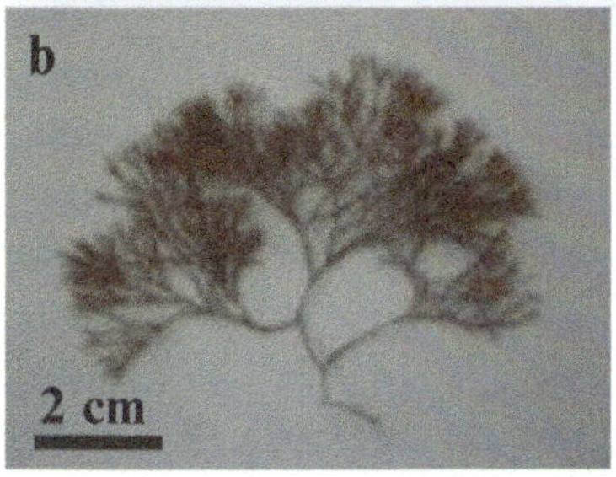

Distribution
Adri: N 20° 57.58'; E 70° 16.76'
Dwarka: N 22° 14.61'; E 68° 57.28'
Okha: N 22° 28.50'; E 69° 04.33'
Porbandar: N 21° 37.92'; E 69° 36.48'
Veraval: N 20° 54.87'; E 70° 20.83'

Jania rubens **(Linnaeus) Lamouroux**

Thallus form and type	:	Calcified clusters
Substratum	:	Intertidal rocks, coralline stones
Occurrence	:	Mid littoral zone, tide pools, some times as epiphyte
Season	:	November–April
Abundance	:	Moderate
Description	:	Plants pinkish red in colour, 2 – 5 cm in height, calcified growing as closely branched tufts; branching dichotomous or dichotomous cymoid, genicula unizonal, segments or intergenicula cylindrical less than 250 μm broad, multizonal, pointed at the apices (Figs. a, b); conceptacles terminal with short antennae and a single large apical pore.

Gigartinales, Caulacanthaceae

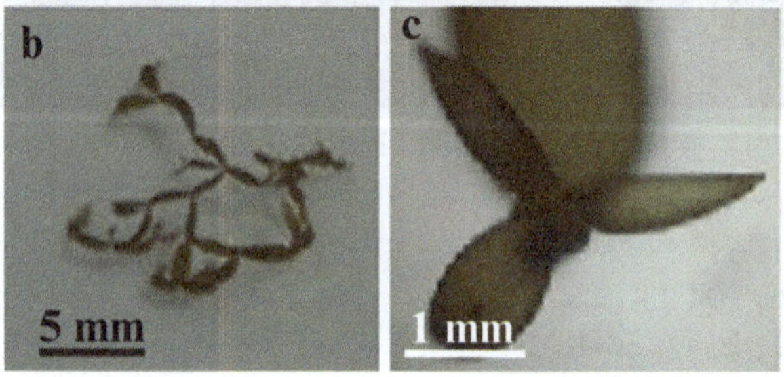

Distribution
Gopnath: N 21° 12.52'; E 72° 06.53'

Catenella impudica **(Montagne) J. Agardh**

Thallus form and type	:	Creeping and articulated spindle shaped segments
Substratum	:	Intertidal rocks
Occurrence	:	Mid littoral zone
Season	:	November – April
Abundance	:	Scanty
Description	:	Plants blackish red in colour, about 1.5 cm high, decumbent, slightly flabellately expanded, articulate with irregularly di- or tri-chotomously branched (Figs. a, b); the young segments slender, the older segments broader and more flattened, segments elliptical – oblong and deeply constricted at the nodes; ultimate segments cylindrical, prolonged into acuminate tips (Fig. c); haptera developing at nodes.
Use	:	A source of carrageenan.

Gigartinales, Gigartinaceae

Distribution
Adri: N 20° 57.58'; E 70° 16.76'
Okha: N 22° 28.42'; E 69° 04.17'

Chondracanthus acicularis **(Roth) Fredericq**

Thallus form and type	:	Filamentous, cylindrical and prostrate
Substratum	:	Intertidal rocks, calcareous substrata
Occurrence	:	Supra littoral zone, pools and crevices in the intertidal
Season	:	July–November
Abundance	:	Scanty
Description	:	Plants dark red in colour, prostrate, cylindrical, slender, 3 – 5 cm in height, 0.5 mm thick, much branched, irregularly dichotomously and alternately branched; branches recurved, sometimes compressed with acuminate apices (Figs. a, b); thallus composed of filamentous medulla and cortex of anticlinal rows of cells.

Gigartinales, Cystocloniaceae

Distribution
Dwarka: N 22° 14.61'; E 68° 57.28'
Okha: N 22° 28.56'; E 69° 04.45'
Porbandar: N 21° 37.92'; E 69° 36.48'
Veraval: N 20° 54.87'; E 70° 20.83'

Hypnea flagelliformis **Greville ex J. Agardh**

Thallus form and type	:	Bushy tufts
Substratum	:	Intertidal rocks
Occurrence	:	Mid littoral zone, tide pools, rock crevices
Season	:	November–April
Abundance	:	Scanty
Description	:	Thallus purplish red in colour, 10 – 15 cm tall with many fronds arising from a common holdfast (Fig. a); fronds sparsely branched, flagelliform, covered with ramuli from base to apex; fertile fronds very much attenuate, slightly thicker in the middle; ramuli densely crowded giving the frond a tomentose appearance, patent to horizontally emergent, simple, subulate.

Gigartinales, Cystocloniaceae

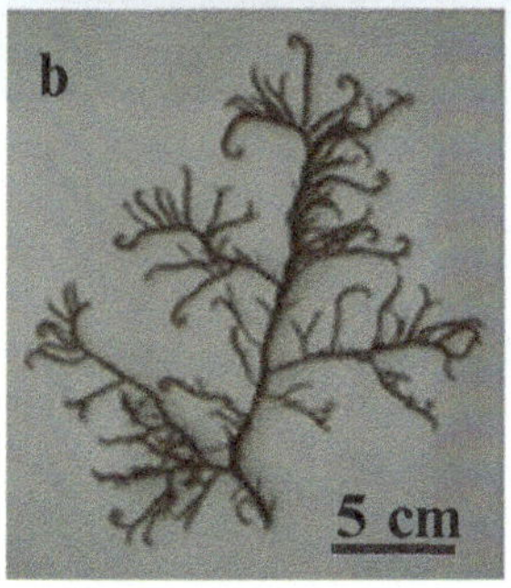

Distribution
Dwarka: N 22° 14.61'; E 68° 57.28'
Okha: N 22° 28.51'; E 69° 04.48'
Porbandar: N 21° 37.92'; E 69° 36.48'
Shivrajpur: N 22° 19.87'; E 68° 56.95'
Veraval: N 20° 54.87'; E 70° 20.83'

Hypnea musciformis **(Wulfen) Lamouroux**

Thallus form and type	:	Bushy with apical hooks
Substratum	:	Intertidal rocks
Occurrence	:	Mid littoral zone, tide pools, epiphyte on larger algae
Season	:	November–April
Abundance	:	Dense
Description	:	Plants pinkish red in colour, up to 30 cm in length, initially attached to substratum, later often free floating, freely branched; branches covered with short spine like branches, apices of the branches often thickened and incurved in the form of hook like structures (Figs. a, b).
Remark	:	All *Hypnea* species are used for food and medicines in China, Japan and other countries and yield k-carrageenan. Cultivation technology (monocline method) has been developed by CSMCRI, Bhavnagar

Gigartinales, Cystocloniaceae

Distribution
Dwarka: N 22° 14.61'; E 68° 57.28'
Okha: N 22° 28.48'; E 69° 04.48'
Porbandar: N 21° 37.92'; E 69° 36.48'
Veraval: N 20° 54.87'; E 70° 20.83'

Hypnea pannosa **J. Agardh**

Thallus form and type	:	Creeping tufts or mats
Substratum	:	Intertidal rocks
Occurrence	:	Lower mid littoral zone infra littoral fringe, rock pools
Season	:	November–April
Abundance	:	Moderate
Description	:	Plants purple to pinkish red in colour, 5 – 10 cm in height, growing as spongy tufts, brittle, easily broken, intricately branched, compressed with conical and acuminate tips (Figs. a, b); fertile branches smooth at the base, pyramidally ramose from the median part towards the apex; carpoginia is on one side of somewhat swollen parts of upper branches.

Gigartinales, Cystocloniaceae

Distribution
Dwarka: N 22° 14.61'; E 68° 57.28'
Okha: N 22° 28.49'; E 69° 04.53'
Porbandar: N 21° 37.92'; E 69° 36.48'
Veraval: N 20° 54.87'; E 70° 20.83'

***Hypnea spinella* (C. Agardh) Kützing**

Thallus form and type	:	Intricate cushions
Substratum	:	Intertidal rocks
Occurrence	:	Lower mid littoral zone infra littoral fringe
Season	:	July–November
Abundance	:	Moderate
Description	:	Plants pinkish to dark red in colour, 2.5 cm in height, growing as cushion or patches; fronds pulvinate, intricately, alternately branched; branches cylindrical, tapering from base to apices, covered with small spines (Figs. a, b).

Gigartinales, Cystocloniaceae

Distribution
Dwarka: N 22° 14.61'; E 68° 57.28'
Okha: N 22° 28.54'; E 69° 04.38'
Porbandar: N 21° 37.92'; E 69° 36.48'
Shivrajpur: N 22° 19.87'; E 68° 56.95'
Veraval: N 20° 54.87'; E 70° 20.83'

Hypnea valentiae **(Turner) Montagne**

Thallus form and type	:	Bushy with spinous ramuli
Substratum	:	Intertidal rocks
Occurrence	:	Lower mid littoral zone, rock pools
Season	:	November–April
Abundance	:	Rare
Description	:	Plants greenish to pinkish red in colour with many axes arising from the basal part, alternately branched, branches caespitose, densely covered with spinous ramuli throughout the length, less densely covered towards the apex; spines given out all round, held horizontally in the lower parts of branches and patent towards the tip, mostly simple, tapering from the base to the acuminate tip, alternately stellately forked (Figs. a, b).

Gigartinales, Cystocloniaceae

Distribution
Dwarka: N 22° 14.61'; E 68° 57.28'
Okha: N 22° 28.50'; E 69° 04.54'
Porbandar: N 21° 37.92'; E 69° 36.48'
Veraval: N 20° 54.87'; E 70° 20.83'

***Hypnea esperi* Bory de Saint-Vincent**

Thallus form and type	:	Remiform with spine like ramuli
Substratum	:	Intertidal rocks, calcareous stones
Occurrence	:	Mid littoral zone
Season	:	November–April
Abundance	:	Moderate
Description	:	Plants greenish red in colour, about 20 cm tall, attached by discoid holdfast; axes 2 – 3 time branched; branches and branchlets closely arranged on the axes, covered with small spines (Figs. a, b).

Gigartinales, Solieriaceae

Cultivated currently at
Okha: N 22° 28.33'; E 69° 04.54'

Kappaphycus alvarezii **(Doty) Doty ex P. Silva**

Thallus form and type	:	Bushy and succulent
Substratum	:	Cultivated material
Occurrence	:	Not present in wild conditions
Season	:	Cultivated throughout the year
Abundance	:	Not applicable
Description	:	Plants dark green in colour, succulent and somewhat stiff, erect and profusely branched with cylindrical axis and branches; branches larger at the bases, small spinous tapering above, axiferous (Figs. a, b); medullary core of axial hyphae persists up to 10 cm below the tip (Fig. c).
Use	:	An important carragenophyte. CSMCRI has developed an integrated method for simultaneous production of nutrient rich liquid sap (liquid fertilizer) and a residue rich of kappa-carrageenan (U. S. Patent No. 6,893,479 dated May 17, 2005). The liquid fertilizer found to be very effective on different crops. Commercially cultivated in India

Gigartinales, Solieriaceae

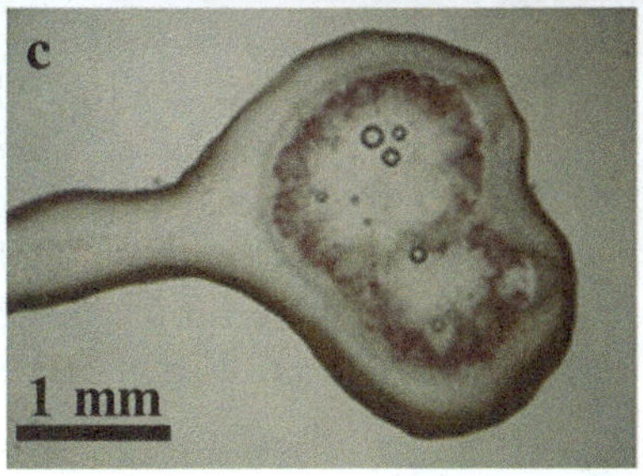

Distribution
Adri: N 20° 57.58'; E 70° 16.76'
Okha: N 20° 28.42'; E 69° 03.57'
Veraval: N 20° 54.87'; E 70° 20.83'

***Meristotheca papulosa* (Montagne) J. Agardh**

Thallus form and type	:	Frondose and fleshy
Substratum	:	Intertidal rocks and coralline stones
Occurrence	:	Infra littoral fringe and sub littoral zone
Season	:	January–April
Abundance	:	Scanty
Description	:	Plants rose red in colour, 5 – 10 cm tall, attached by a discoid or crustose holdfast; flattened, foliose, strap-like and fleshy with pinnately or palmately dissected or regularly lobed thalli (Figs. a, b); ultimate laterals terete and tapering to tips; thallus is multiaxial with cortical layer and filamentous medulla; cystocarps are sunk in the cortex surrounded by filamentous pericarp, carpospores ellipsoidal (Fig. c).
Use	:	Popularly used in salad preparations in Japan. Japanese have made some attempts to develop field cultivation methods for large scale propagation like other seaweeds

Gigartinales, Solieriaceae

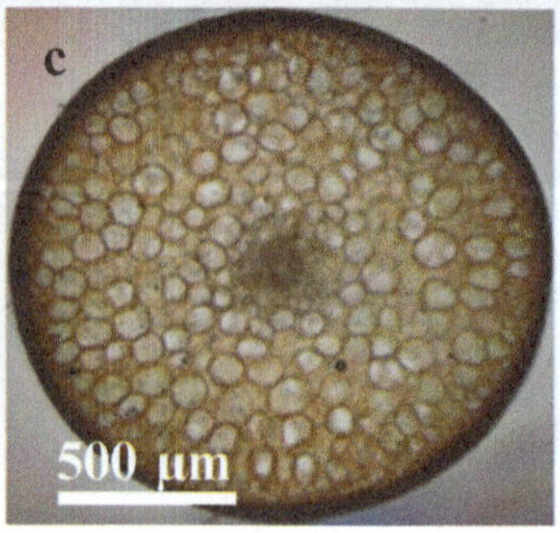

Distribution
Dwarka: N 22° 14.61'; E 68° 57.28'
Okha: N 22° 28.45'; E 69° 03.59'
Okha: N 22° 26.38'; E 69° 03.30'
Porbandar: N 21° 37.92'; E 69° 36.48'
Shivrajpur: N 22° 19.87'; E 68° 56.95'
Veraval: N 20° 54.87'; E 70° 20.83'

***Sarconema filiforme* (Sonder) Kylin**

Thallus form and type	:	Remiform and cylindrical
Substratum	:	Intertidal rocks and calcareous stones
Occurrence	:	Lower mid littoral zone, rock pools
Season	:	November–April
Abundance	:	Moderate
Description	:	Plants purplish red or brownish red in colour, slender and filamentous, up to 15 cm in height, solid, terete with cylindrical axes, regularly and dichotomously branched, tapering gradually towards the tip (Figs. a, b); thallus composed of rectangular peripheral cells, and paranchymatous cortex and a central strands and densely packed thick walled cells (Fig. c).

Gigartinales, Solieriaceae

Distribution
Adri: N 20° 57.58'; E 70° 16.76'
Dwarka: N 22° 14.61'; E 68° 57.28'
Okha: N 22° 28.42'; E 69° 03.58'
Porbandar: N 21° 37.92'; E 69° 36.48'
Shivrajpur: N 22° 19.87'; E 68° 56.95'
Veraval: N 20° 54.87'; E 70° 20.83'

***Sarconema scinaioides* Børgesen**

Thallus form and type	:	Remiform and cylindrical
Substratum	:	Intertidal rocks, coralline stones
Occurrence	:	Mid littoral zone, tide pools
Season	:	November–April
Abundance	:	Dense
Description	:	Plants copper red in colour, fleshy, more than 14 cm tall, erect, caespitose, terete, repeatedly, irregularly and dichotomously branched; fronds about 2 mm in diameter, terminal portions gradually tapering to acute tips (Figs. a, b); thallus composed of an epidermal layer of uniform cells and large cortical cells; cuticle thick; the central medullary filaments made up of thick filaments.

Gigartinales, Solieriaceae

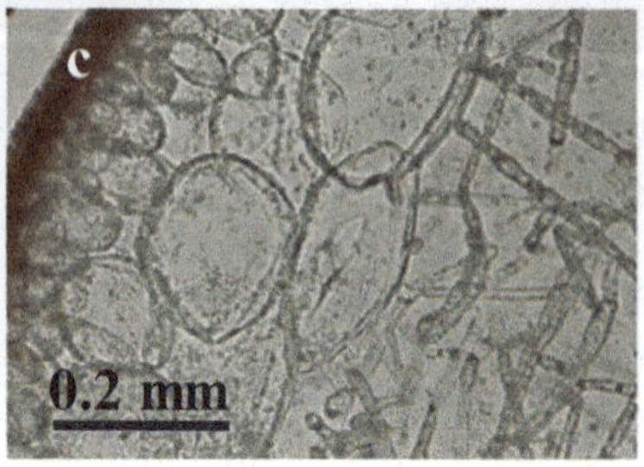

Distribution
Porbandar: N 21° 37.92';
E 69° 36.48'

***Solieria chordalis* (C. Agardh) J. Agardh**

Thallus form and type	:	Bushy and succulent
Substratum	:	Intertidal rocks, coralline stones
Occurrence	:	Lower mid littoral zone, tide pools
Season	:	November–April
Abundance	:	Scanty
Description	:	Plants bright red in colour, up to 15 cm high attached by a small discs; many axes arising from discoid holdfast; axes and main branches profusely branched, 2 – 4 mm broad; branching alternate, sub-dichotomous or whorled with fusiform and acute ramuli (Figs. a, b); thallus composed of 1 – 2 layered epidermis, large celled cortex and filamentous or hyphae-like medulla (Fig. c).

Gigartinales, Solieriaceae

Distribution
Dwarka: N 22° 14.61'; E 68° 57.28'
Kotda: N 20° 42.38'; E 70° 52.17'
Okha: N 22° 28.41'; E 69° 03.57'
Porbandar: N 21° 37.92'; E 69° 36.48'
Shivrajpur: N 22° 19.87'; E 68° 56.95'
Veraval: N 20° 54.87'; E 70° 20.83'

***Solieria robusta* (Greville) Kylin**

Thallus form and type	:	Bushy and succulent
Substratum	:	Intertidal rocks, coralline stones
Occurrence	:	Mid littoral zone
Season	:	November–April
Abundance	:	Moderate
Description	:	Plants dark pink to brownish red in colour, more than 15 cm in height, fleshy, profusely branched (Figs. a, b); branches terete or slightly compressed, 1 – 5 mm broad, basally constricted with sub-acute tips; branching irregular from all sides, sometimes umbellate with 4 – 5 branches arising from proliferously from a damaged branch.

Nemastomatales, Nemastomataceae

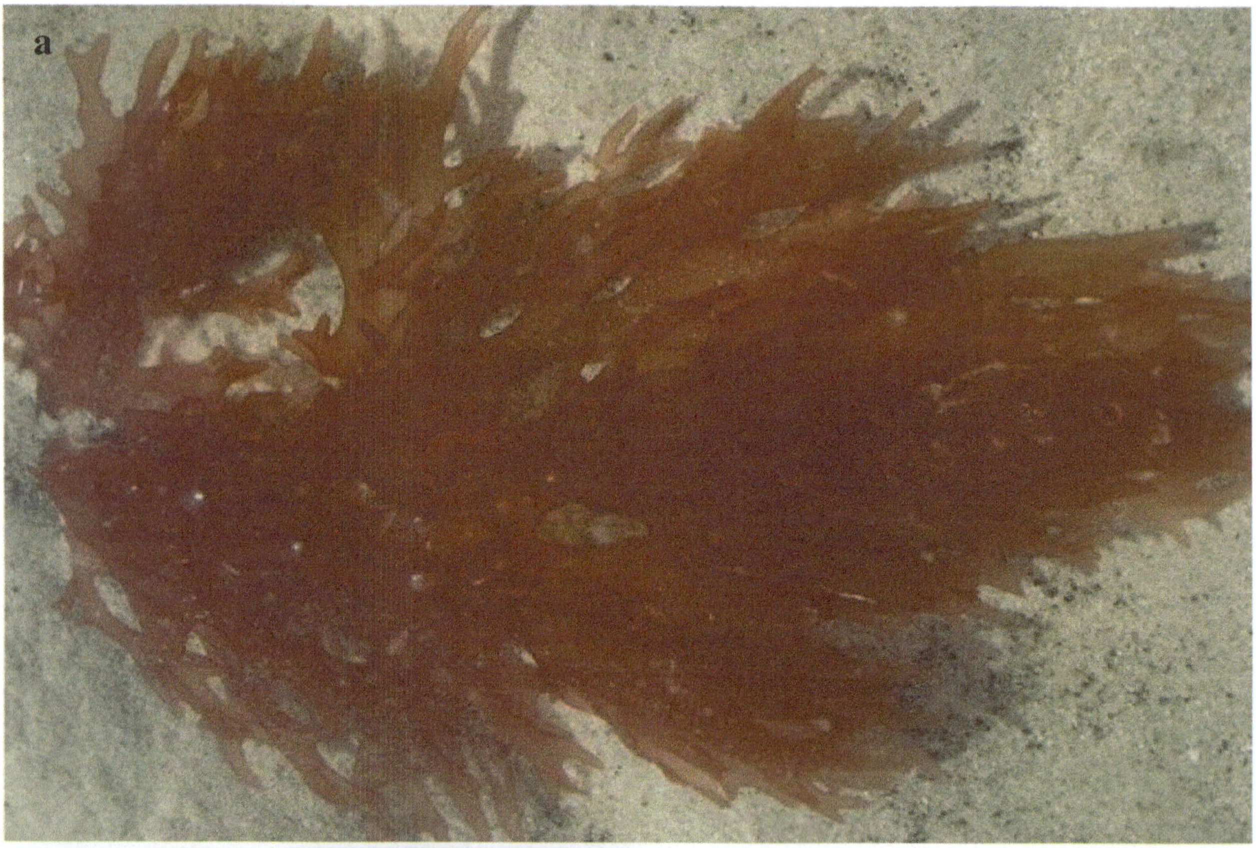

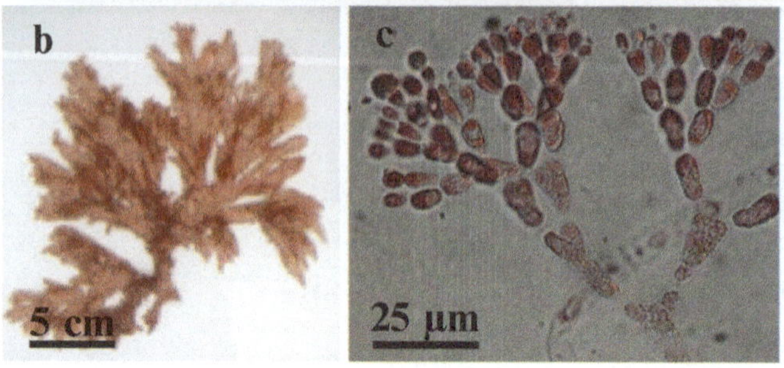

Distribution
Okha: N 22° 28.45'; E 69° 04.51'

Predaea feldmannii **Børgesen var. *indica* M. Balakrishnan & Chawla**

Thallus form and type	:	Mucilaginous and sinuate
Substratum	:	Bed rocks
Occurrence	:	Sub littoral zone
Season	:	February–April
Abundance	:	Scanty
Description	:	Plants yellowish red in colour, 10 – 20 cm tall, soft, mucilaginous, irregular with finger shaped or wart like processes, complanate to terete; margins of the thallus sinuate and laciniate (Figs. a, b); thallus multiaxial composed of filamentous medulla and a cortical layer of dichotomously branched filaments (Fig. c); carpogonial branches 2 – 4 celled.

Sebdeniales, Sebdeniaceae

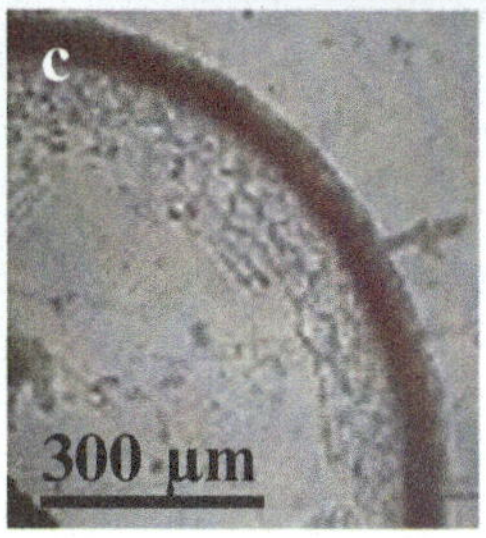

Distribution
Dwarka: N 22° 14.61'; E 68° 57.28'
Okha: N 22° 28.43'; E 69° 04.51'
Porbandar: N 21° 37.92'; E 69° 36.48'
Veraval: N 20° 54.87'; E 70° 20.83'

Sebdenia flabellata **(J. Agardh) Parkinson**

Thallus form and type	:	Soft and succulent
Substratum	:	Intertidal rocks, rock pools
Occurrence	:	Mid littoral zone and infra littoral fringe
Season	:	November–April
Abundance	:	Moderate
Description	:	Plants pinkish red in colour, 25 – 30 cm tall, cylindrical, soft, developing from a cuneate base, repeatedly dichotomous (Figs. a, b), showing multiaxial construction; segments 5 – 10 mm broad; medulla filamentous without a distinct central axis; cortex densely crowded with inner large cells and outer small cells (Fig. c).

Rhodymeniales, Champiaceae

Distribution
Dwarka: N 22° 14.61'; E 68° 57.28'
Kotda: N 20° 42.38'; E 70° 52.17'
Okha: N 22° 28.45'; E 69° 03.00'
Porbandar: N 21° 37.92'; E 69° 36.48'
Shivrajpur: N 22° 19.87'; E 68° 56.95'
Veraval: N 20° 54.87'; E 70° 20.83'

Champia compressa **Harvey**

Thallus form and type	:	Soft, segmented and mucilaginous
Substratum	:	Intertidal rocks, coralline stones
Occurrence	:	Lower mid littoral zone, tide pools
Season	:	November–April
Abundance	:	Dense
Description	:	Plants purple red to blackish red in colour, up to 5 cm tall, gelatinous, creeping; fronds compressed to flattened, 2 – 4 mm wide articulate, branching, sub-distichous bi-pinnate, branches markedly narrow at the base expended above, segments short barrel-shaped, 0.8 – 1 mm wide (Figs. a, b).

Rhodymeniales, Champiaceae

Distribution
Okha: N 22° 28.45'; E 69° 04.00'
Porbandar: N 21° 37.92'; E 69° 36.48'
Shivrajpur: N 22° 19.87'; E 68° 56.95'

***Champia globulifera* Børgesen**

Thallus form and type	:	Soft, segmented and mucilaginous
Substratum	:	Intertidal rocks, coralline stones
Occurrence	:	Lower mid littoral zone and infra littoral fringe
Season	:	November–April
Abundance	:	Dense
Description	:	Plants dark red to violet red in colour, 4 – 6 cm tall, segmented, densely caespitose, gelatinous, attached by rhizoids and basal discs, sometimes decumbent branches produced; branching monopodial, opposite, verticillate or alternate, simple or branched, broad in the middle and tapering gradually near the apices, apices obtuse; segments barrel-shaped, 2 mm broad as long as broad (Figs. a, b).

Rhodymeniales, Champiaceae

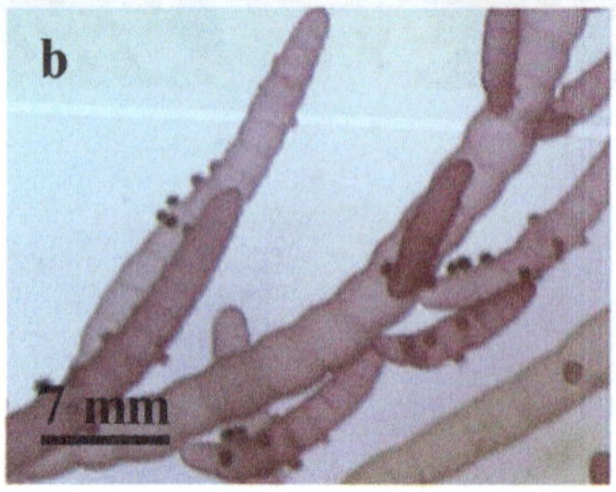

Distribution
Dwarka: N 22° 14.61'; E 68° 57.28'
Okha: N 22° 28.41'; E 69° 03.58'
Porbandar: N 21° 37.92'; E 69° 36.48'
Shivrajpur: N 22° 19.87'; E 68° 56.95'
Veraval: N 20° 54.87'; E 70° 20.83'

***Champia indica* Børgesen**

Thallus form and type	:	Soft, segmented and mucilaginous
Substratum	:	Intertidal rocks, calcareous substrata
Occurrence	:	Lower mid littoral zone and infra littoral fringe
Season	:	November–April
Abundance	:	Scanty
Description	:	Plants dark red to pinkish red in colour, up to 15 cm tall, segmented, gelatinous, densely caespitose and growing as pyramidal tufts (Figs. a, b); articulations pronounced with barrel shaped segments, segments 3 mm broad, as long as broad and 1.5 mm long in the side branches; cystocarps small, sub-spherical and scattered over the surface of the segments.

Rhodymeniales, Champiaceae

Distribution
Okha: N 22° 28.41'; E 69° 03.58'
Porbandar: N 21° 37.92'; E 69° 36.48'
Shivrajpur: N 22° 19.87'; E 68° 56.95'
Veraval: N 20° 54.87'; E 70° 20.83'

***Champia parvula* (C. Agardh) Harvey**

Thallus form and type	:	Soft, segmented and mucilaginous
Substratum	:	Intertidal rocks
Occurrence	:	Lower mid littoral zone, epiphytic on larger algae
Season	:	November–April
Abundance	:	Scanty
Description	:	Plants crimson red in colour (Fig. a), 3 – 7 cm tall, gelatinous, segmented, generally epiphytic in nature, caespitose and attached by the rhizoids; thallus closely branched with decumbent and erect branches with obtuse tips, few decumbent branches usually present at the base; segments barrel shaped, slightly constricted near the septa, as long as broad or much longer.

Rhodymeniales, Champiaceae

Distribution
Dwarka: N 22° 14.61'; E 68° 57.28'
Kotda: N 20° 42.18'; E 70° 26.25'
Okha: N 22° 28.43'; E 69° 03.59'
Porbandar: N 21° 37.92'; E 69° 36.48'
Veraval: N 20° 54.87'; E 70° 20.83'

***Champia somalensis* Hauk**

Thallus form and type	:	Cartilaginous, segmented and mucilaginous
Substratum	:	Intertidal rocks
Occurrence	:	Lower mid littoral zone, rock pools
Season	:	November–April
Abundance	:	Moderate
Description	:	Plants dark red in colour, up to 20 cm high, tough and cartilaginous, segmented, caespitose, growing as pyramidal tufts, attached by rhizoids; basal half devoid of branches; branching irregularly pinnate, alternate or opposite, sub-distichous, ultimate branches clavate; segments 2 – 3 mm wide, shorter than broad (Figs. a, b).

Rhodymeniales, Champiaceae

Distribution
Dwarka: N 22° 14.61'; E 68° 57.28'
Okha: N 22° 28.43'; E 69° 03.59'
Porbandar: N 21° 37.92'; E 69° 36.48'
Veraval: N 20° 54.87'; E 70° 20.83'

Gastroclonium iyengarii **K. Srinivasan**

Thallus form and type	:	Bushy with hollow segmented branch
Substratum	:	Intertidal rocks, calcareous substrata
Occurrence	:	Lower mid littoral zone and infra littoral fringe
Season	:	November–April
Abundance	:	Scanty
Description	:	Plants dark brown colour in older parts and light green to light red at the upper parts (Figs. a, b), up to 20 cm tall, cartilaginous, erect and cylindrical; thallus pseudo dichotomously branched, naked and solid below and bearing a number of short inflated segmented ramuli at the upper parts of the branches; ramuli pink to red in younger stage and light red to purple with shades of green in older stage, laterally disposed in acropetal succession, inflated, linear-oblong, segmented with a narrow base and acute apex (Figs. a, b); segments of the middle portion of the ramuli are bigger than the distal portion, 15 – 20 mm long, 2.5 – 3.5 mm broad.

Rhodymeniales, Rhodymeniaceae

Distribution
Dwarka: N 22° 14.61'; E 68° 57.28'
Okha: N 22° 28.49'; E 69° 04.56'
Porbandar: N 21° 37.92'; E 69° 36.48'
Shivrajpur: N 22° 19.87'; E 68° 56.95'
Veraval: N 20° 54.87'; E 70° 20.83'

Botryocladia leptopoda **(J. Agardh) Kylin**

Thallus form and type	:	Bushy with mucilaginous spherical vesicles
Substratum	:	Intertidal rocks
Occurrence	:	Infra littoral fringe and sub littoral zone, rock crevices
Season	:	November–April
Abundance	:	Moderate
Description	:	Plants pinkish red to violet red in colour, up to 60 cm tall, erect, attached by discoid holdfasts; stems cylindrical, thick, densely and irregularly branched, solid and covered with sub-spherical and minutely stalked vesicles of 1 – 5 mm diameter (Figs. a, b).

Rhodymeniales, Rhodymeniaceae

Distribution
Okha: N 22° 28.47'; E 69° 04.55'
Porbandar: N 21° 37.92'; E 69° 36.48'
Sutrapada: N 20° 50.18'; E 70° 28.89'
Veraval: N 20° 54.87'; E 70° 20.83'

***Coelarthrum muelleri* (Sonder) Børgesen**

Thallus form and type	:	Delicate and ovular vesicles
Substratum	:	Intertidal rocks
Occurrence	:	Infra littoral fringe, rock pools
Season	:	November–April
Abundance	:	Scanty
Description	:	Plants bright red in colour, up to 25 cm tall, erect, attached by a solid creeping stoloniferous basal portion with 2 – 3 mm thick branches; distinctly articulated regularly di- or trichotomously branched; segments cylindrical, ovoid or cuneate, hollow; inter-segmentary portions solid and cylindrical (Figs. a, b).

Rhodymeniales, Rhodymeniaceae

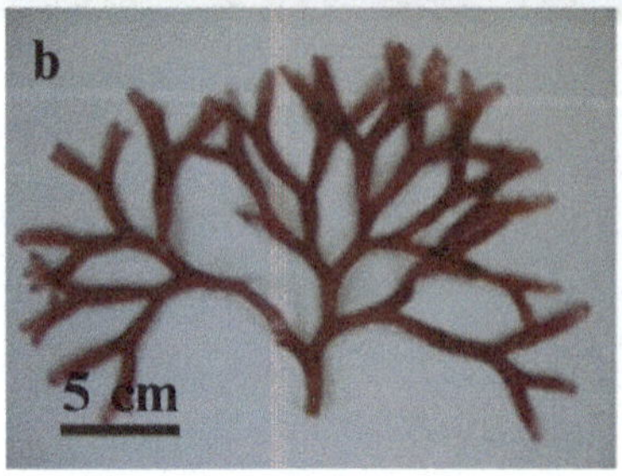

Distribution
Adri: N 20° 57.58'; E 70° 16.76'
Dwarka: N 22° 14.61'; E 68° 57.28'
Okha: N 22° 28.43'; E 69° 04.50'
Porbandar: N 21° 37.92'; E 69° 36.48'
Veraval: N 20° 54.87'; E 70° 20.83'

***Rhodymenia sonderi* P. Silva**

Thallus form and type	:	Ribbon like with cylindrical stalks
Substratum	:	Intertidal rocks
Occurrence	:	Infra littoral fringe, rock pools
Season	:	November–April
Abundance	:	Scanty
Description	:	Plants dark red or rose pink in colour, up to 15 cm tall, erect, firm and gelatinous and attached by a distinct basal disc; thallus dichotomously divided; blades stalked, 2 – 3 mm wide, lower portions cuneate, upper portions linear, sub-flabellately or fastigiately branched, angle between furcations acute (Figs. a, b); margins of the blade entire; fronds slightly thickened at the margins and more or less sub-canaliculate or concave on one side.

Rhodymeniales, Lomantariaceae

Distribution
Shivrajpur: N 22° 19.87'; E 68° 56.95'

***Gelidiopsis repens* (Kützing) Shmitz**

Thallus form and type	:	Filiform, cylindrical and cushion
Substratum	:	Intertidal rocks
Occurrence	:	Mid littoral zone, rock pools
Season	:	July–November
Abundance	:	Scanty
Description	:	Plants light red in colour (Fig. a), 3 – 4 cm in height, filiform, dichotomously alternately and sometimes pinnately branched; branches spread out in subflagellate manner, upper segments slender, the thallus composed of central medulla of small cells, cortex with large cells and a surface layer of small cells.

Rhodymeniales, Lomentariaceae

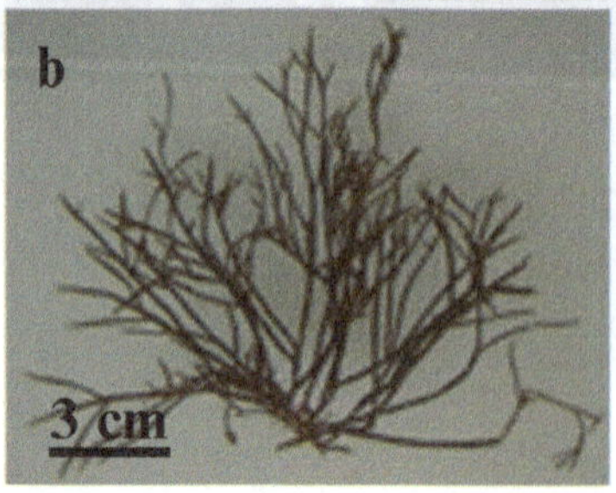

Distribution
Adri: N 20° 57.58'; E 70° 16.76'
Okha: N 22° 28.52'; E 69° 04.47'
Porbandar: N 21° 37.92'; E 69° 36.48'
Veraval: N 20° 54.87'; E 70° 20.83'

***Gelidiopsis variabilis* (J. Agardh) Schmitz**

Thallus form and type	:	Bushy and cartilagenous
Substratum	:	Intertidal rocks
Occurrence	:	Mid littoral zone, rocks bordering tide pool
Season	:	November–April
Abundance	:	Scanty
Description	:	Plants brownish red in colour, 7 – 10 cm high, primary axis erect, cylindrical, filiform, wiry sparingly branched and stiff; branches simple, cylindrical with obtuse apices (Figs. a, b).
Use	:	This alga is used in human food and medicine

Ceramiales, Wrangeliaceae

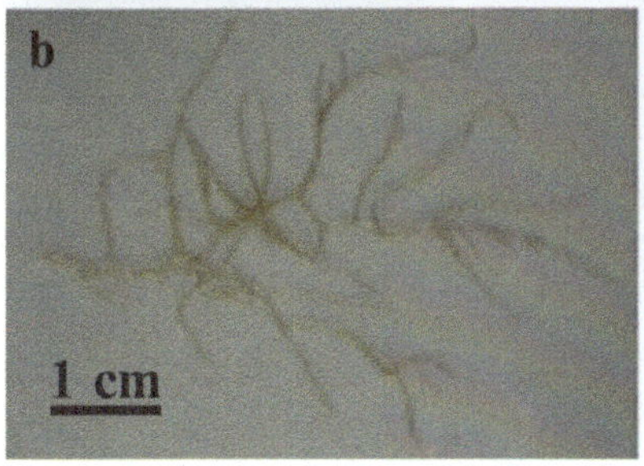

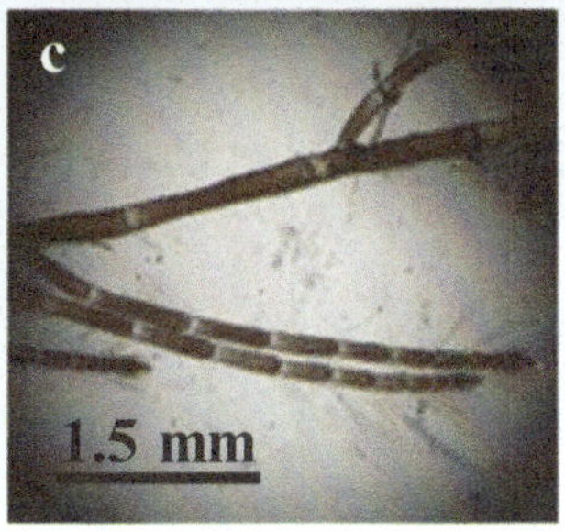

Distribution
Adri: N 20° 57.58'; E 70° 16.76'
Dwarka: N 22° 14.61'; E 68° 57.28'
Okha: N 22° 26.38'; E 69° 03.30'
Porbandar: N 21° 38.24'; E 69° 35.81'
Veraval: N 20° 54.87'; E 70° 20.83'

***Anotrichium tenue* (C. Agardh) Nägeli**

Thallus form and type	:	Remiform and bushy
Substratum	:	Intertidal rocks, rock pools
Occurrence	:	Mid littoral zone, epiphytic on larger algae
Season	:	November – April
Abundance	:	Scanty
Description	:	Plants light red in colour, 2 – 5 cm tall or more, erect; filaments sparsely branched with long and short branches, branching divaricate or branches arising at right angles to the axes; apices of the branches crowned with a whorl of forking branchlets, cells somewhat cylindrical, constricted at the joints (Figs. a – c).

Ceramiales, Wrangeliaceae

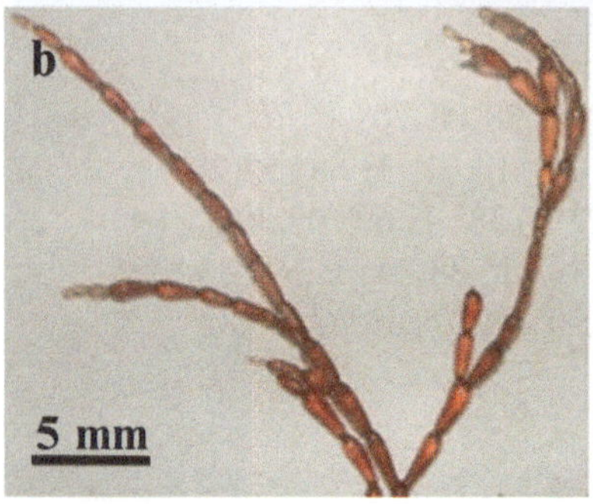

Distribution
Okha: N 22° 28.44'; E 69° 04.49'
Porbandar: N 21° 37.92'; E 69° 36.48'
Sutrapada: N 20° 50.18'; E 70° 28.89'
Veraval: N 20° 54.87'; E 70° 20.83'

Griffithsia corallinoides **(Linnaeus) Trevisan**

Thallus form and type	:	Bushy and beaded
Substratum	:	Intertidal rocks
Occurrence	:	Infra littoral fringe, rock pools
Season	:	November–April
Abundance	:	Scanty
Description	:	Plants shiny red to pinkish red in colour, 6 – 12 cm tall, erect, bushy with dichotomous and fastigiate branching; upper cells moniliform, oblong, 2 – 5 times as long as broad; cells in the lower parts of the frond sub-cylindrical, constricted at the junction between superposed cells, 3 – 4 times longer than broad (Figs. a, b); tetrasporangia in whorls of fascicled clusters.

Ceramiales, Wrangeliaceae

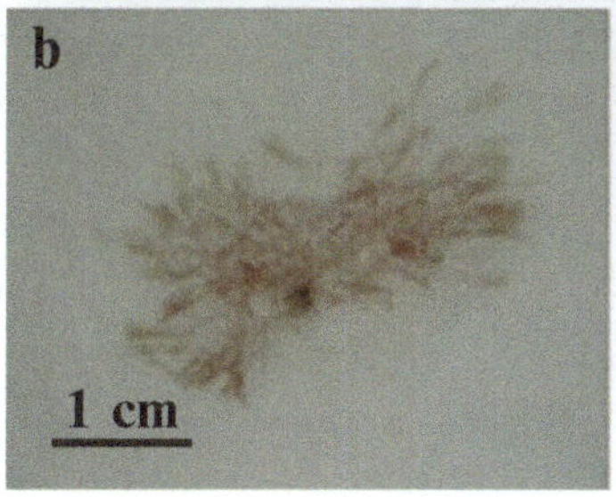

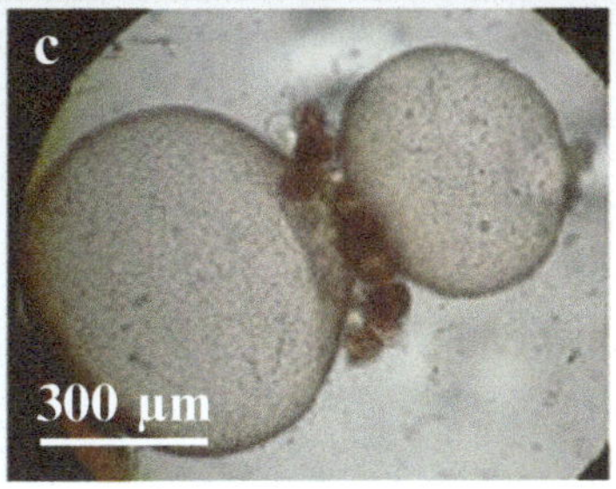

Distribution
Okha: N 22° 28.48'; E 69° 04.56'
Porbandar: N 21° 37.92'; E 69° 36.48'
Shivrajpur: N 22° 19.87'; E 68° 56.95'

Griffithsia opuntioides **J. Agardh**

Thallus form and type	:	Creeping and beaded
Substratum	:	Intertidal rocks
Occurrence	:	Lower mid littoral zone and infra littoral fringe
Season	:	November–April
Abundance	:	Scanty
Description	:	Plants light red to brownish red in colour, 2 – 5 cm tall, mucilaginous in texture, regularly dichotomous, growing in patches, fastigiate, flabellate with claviform cells (Figs. a, b); tetrasporangia in verticilate, vesicles surrounded by a ring of involucral cells (Fig. c).

Ceramiales, Wraneliaceae

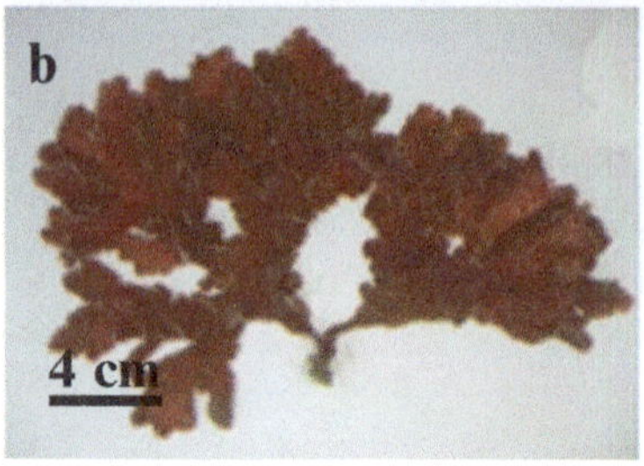

Distribution
Adri: N 20° 57.58'; E 70° 16.76'
Okha: N 22° 28.36'; E 69° 04.54'
Veraval: N 20° 54.87'; E 70° 20.83'

***Haloplegma duperreyi* Montagne**

Thallus form and type	:	Leafy with spongy texture
Substratum	:	Intertidal rocks, corallines tones
Occurrence	:	Infra littoral fringe and sub littoral zone
Season	:	November–April
Abundance	:	Scanty
Description	:	Plants dark red in colour, up to 12 cm in height, spongy with reticulately joined filaments (Figs. a, b), stipitate below and expended into flabelliform fronds above; fronds distinctly lobed and irregularly divided into the segments; segments up to 2 cm broad, lobes oblong with minutely fimbriate margin.

Ceramiales, Wrangeliaceae

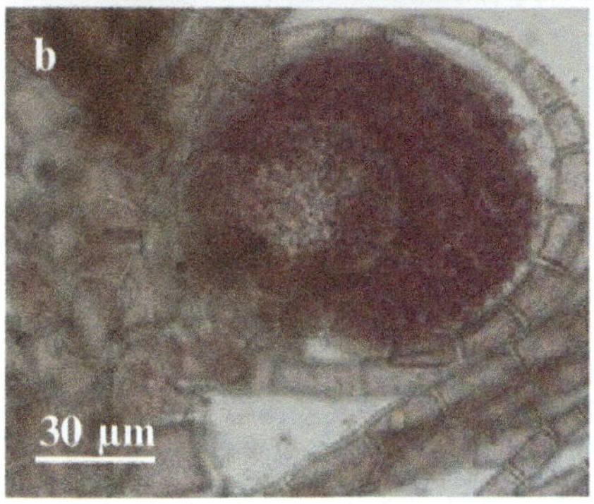

Distribution
Okha: N 22° 28.50'; E 69° 04.52'

Pleonosporium polymorphum **Itono**

Thallus form and type	:	Filamentous and bushy
Substratum	:	Intertidal rocks
Occurrence	:	Infra littoral fringe and sub littoral zone
Season	:	July–November
Abundance	:	Scanty
Description	:	Plants pink dark colour, 5 – 6 cm tall, epiphytic, attached by many rhizoids arising from the basal parts of the erect axes (Fig. a); erect axis about 120 μm or more thick at the basal parts and thinner above; branching alternate, distichous, single un-branched filaments arising in the lower cells of the branches, these un-branched filaments growing downwards; cystocarps globular in shape, naked or with rudimentary involucral filaments (Fig. b).

Ceramiales, Wrangeliaceae

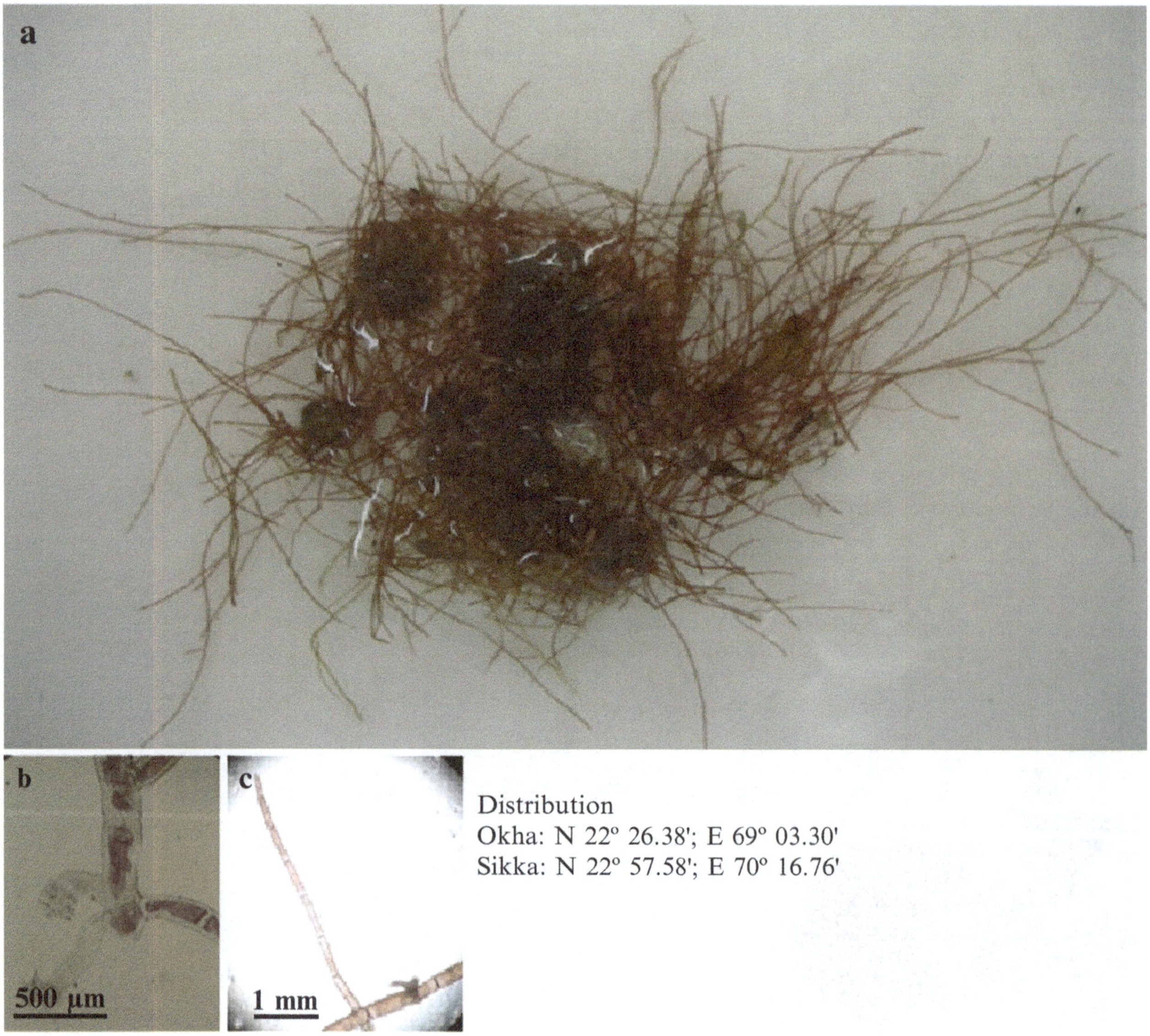

Distribution
Okha: N 22° 26.38'; E 69° 03.30'
Sikka: N 22° 57.58'; E 70° 16.76'

***Spermothamnion* sp**

Thallus form and type	:	Filamentous clumps
Substratum	:	Intertidal rocks, stones, pebbles or shells
Occurrence	:	Mid littoral zone
Season	:	November–April
Abundance	:	Scanty
Description	:	Plants dark red in colour, up to 5 cm in height, forming tufts, filamentous and freely branched with prostrate and erect filaments. Filaments uniseriate, uncorticated, prostrate filaments attached by means of rhizoidal cells ending with discs (Figs. a, b); Erect filaments branched with lateral and opposite branches (Fig. c).

Ceramiales, Wrangeliaceae

Distribution
Adri: N 20° 57.58'; E 70° 16.76'
Navabander: N 22° 45.33'; E 71° 65.23'
Okha: N 22° 28.49'; E 69° 04.24'

Wrangelia tanegana **Harvey**

Thallus form and type	:	Soft and remiform tufts
Substratum	:	Intertidal rocks
Occurrence	:	Mid littoral zone, rock pools
Season	:	November–April
Abundance	:	Moderate
Description	:	Plants dark red in colour, up to 10 cm tall, attached by heptera arising from basal cells of determinate branches on the prostrate axis, thallus bushy with uniseriate axes and four determinate branches at each node in the form of whorl, basal cells of determinate braches developing ramified filaments with cover the upper parts of the axial cells (Figs. a, b).

Ceramiales, Ceramiaceae

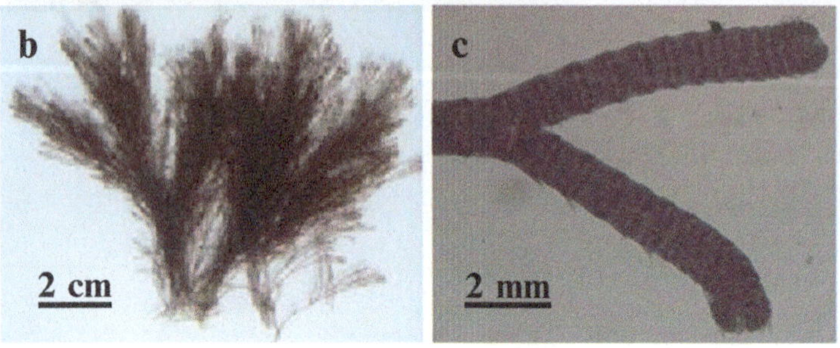

Distribution
Dwarka: N 22° 14.61'; E 68° 57.28'
Kotda: 20° 42.15'; E 70° 52.07'
Okha: N 22° 28.52'; E 69° 04.32'
Porbandar: N 21° 37.92'; E 69° 36.48'
Sutrapada: N 20° 50.18'; E 70° 28.89'
Veraval: N 20° 54.87'; E 70° 20.83'

Centroceras clavulatum **(C. Agardh) Montagne**

Thallus form and type	:	Filamentous and bushy
Substratum	:	Intertidal rocks, calcareous stones
Occurrence	:	Lower mid littoral zone, rock pools
Season	:	November–April
Abundance	:	Moderate
Description	:	Plants dark red in colour, 5 – 8 cm tall, erect, filamentous and rigid; filaments regularly dichotomously branched (Figs. a-c); filaments with nodes and fully corticated internodes, 500 µm long, 120 – 180 µm broad; ultimate branches forcipate, slightly incurved; cortical cells quadrate or rectangular arranged in longitudinal rows, nodes with a ring of 1 – 3 celled spines.

Ceramiales, Ceramiaceae

Distribution
Gopnath: N 21° 12.52';
E 72° 06.53'
Okha: N 22° 28.36';
E 69° 04.50'

***Ceramium cruciatum* Collins & Harvey**

Thallus form and type	:	Filamentous and branched
Substratum	:	Intertidal rocks, attached to structures in the sea
Occurrence	:	Lower mid littoral zone, epiphytic on the larger algae
Season	:	November–April
Abundance	:	Scanty
Description	:	Plants dark red in colour, 2 – 3 cm tall, filamentous with creeping and erect axes; creeping axis attached by colourless unicellular rhizoids, arising fronds in the nodal cortical cells, erect axis sub-dichotomously branched in one plane (Figs. a, b); number of segments between the dichotomous 2 – 4; tetrasporangia produced on the upper parts of the filaments (Fig. c).
Use	:	Largest species are used for food and sometimes mixed with agarophytes for phycocolloid production

Ceramiales, Spyridiaceae

Distribution
Adri: N 20° 57.58'; E 70° 16.76'
Okha: N 22° 28.53'; E 69° 04.40'

Spyridia alternans **Børgesen**

Thallus form and type	:	Remiform with uniseriate ramuli
Substratum	:	Intertidal rocks, calcareous substrata
Occurrence	:	Sub littoral zone, rock pools
Season	:	November–April
Abundance	:	Scanty
Description	:	Plants bright red in colour, up to 25 cm or more in height, caespitose, segmented, irregularly branched, articulate and corticated in the lower parts (Figs. a, b); axes 1.5 mm broad, gradually attenuated towards the apex, branches compressed in the upper parts with short branchlets; branchlets or ramuli uniseriate, distichously arranged, corticated near the septa, about 300 μm long 20 – 50 μm broad, ending with 2 – 3 spines.

Ceramiales, Spyridiaceae

Distribution
Okha: N 22° 26.38'; E 69° 03.30'
Veraval: N 20° 54.87'; E 70° 20.83'

Spyridia filamentosa **(Wulfen) Harvey**

Thallus form and type	:	Remiform with uniseriate ramuli
Substratum	:	Intertidal rocks, stones or pebbles
Occurrence	:	Mid littoral zone
Season	:	November–April
Abundance	:	Scanty
Description	:	Plants dark red to brownish red in colour, 15 – 20 cm tall, bushy, attached by a rhizoidal disc-like holdfast (Figs. a, b); branching alternate, both main axes and branches completely corticated, branches covered with short uniseriate deciduous ramuli (Fig. c); ramuli radially arranged, 0.5 – 1.5 mm long, 20 – 45 µm broad; nodes surrounded by small cortical cells.

Ceramiales, Spyridiaceae

Distribution
Shivrajpur: N 22° 19.87'; E 68° 56.95'

***Spyridia fusiformis* Börgesen**

Thallus form and type	:	Cylindrical and sub-moniliformis
Substratum	:	Intertidal rocks
Occurrence	:	Mid littoral zone, rock crevices
Season	:	July–November
Abundance	:	Scanty
Description	:	Plants whitish pink in colour (Fig. a), 20 cm in height, terete, sub-moniliform, irregularly and alternately branched; axes about 2 mm in width, gradually tapering towards the branches; branches sparse, short, fusiform, corticated; branchlets sparse, filiform, deciduous, 1 mm long and 25 µm broad, corticated at the nodes by a ring of cells, ending in 2 – 3 small sub-conical cells.

Ceramiales, Spyridiaceae

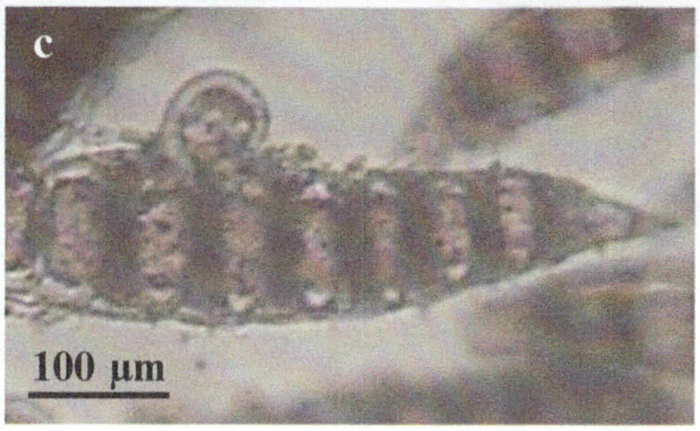

Distribution
Porbandar:
N 21° 38.24';
E 69° 35.81'

Spyridia hypnoides **(Bory de Saint – Vincent) Papenfuss**

Thallus form and type	:	Remiform and feather-like
Substratum	:	Intertidal rocks
Occurrence	:	Lower mid littoral zone
Season	:	November–April
Abundance	:	Scanty
Description	:	Plants dark red in colour, 10 – 15 cm tall, erect, filamentous, densely ramified, segmented and branched; branches and branchlets arranged distichously, giving a feather-like appearance to the frond (Figs. a, b), the main axes and branches completely corticated; ramuli 1 mm long, uniseriate, corticated only at the nodes, 35 – 100 µm broad, tapering above (Fig. c), ending with mucronate tip or with two or three curved spine cells.

Ceramiales, Dasyaceae

Distribution
Okha: N 22° 28.42'; E 69° 03.58'
Porbandar: N 21° 37.92'; E 69° 36.48'
Veraval: N 20° 54.87'; E 70° 20.83'

Dasya punicea **Meneghini ex Zanardini**

Thallus form and type	:	Bushy and polysiphonous
Substratum	:	Intertidal rocks, coralline stones
Occurrence	:	Lower mid littoral zone and infra littoral fringe
Season	:	November–April
Abundance	:	Scanty
Description	:	Plants dark red to pinkish red in colour, 5 – 10 cm tall, attached by a rhizoidal base; main axes multifariously branched giving a pyramidal appearance; upper parts of the axes and branches covered with soft pseudolaterals which occur in whorls (Figs. a, b); basal regions totally devoid of pseudolaterals, axes and branches corticated right up to the tip, pseudolaterals 2 – 3 mm long, dichotomously branched fasciculate.

Ceramiales, Dasyaceae

Distribution
Okha: N 22° 28.49'; E 69° 04.57'
Porbandar: N 21° 37.92'; E 69° 36.48'
Veraval: N 20° 54.87'; E 70° 20.83'

***Heterosiphonia crispella* (C. Agardh) Wynne**

Thallus form and type	:	Remiform and polysiphonous
Substratum	:	Intertidal rocks
Occurrence	:	Upper mid littoral zone
Season	:	November–April
Abundance	:	Scanty
Description	:	Plants dark red in colour (Fig. a), 2 – 3 cm tall, attached by a rhizoidal base; main axis polysiphonous, irregularly branched with 4 – 5 pericentrals, cortication absent; branchlets monosiphonous, sub-dichotomous, basal region of branches polysiphonous and continuous with the main axis; tetrasporangia in stichidia borne on two or three celled monosiphonous stalks; tetrasporangia single in each fertile segment.

Ceramiales, Dasyaceae

Distribution
Okha: N 22° 28.49'; E 69° 04.57'
Porbandar: N 21° 37.92'; E 69° 36.48'
Veraval: N 20° 54.87'; E 70° 20.83'

Heterosiphonia muelleri **(Sonder) De Toni**

Thallus form and type	:	Remiform, polysiphonous and feathery
Substratum	:	Intertidal rocks, calcareous stones
Occurrence	:	Lower mid littoral zone
Season	:	November–April
Abundance	:	Scanty
Description	:	Plants dark red to brownish red in colour, bushy, attached by rhizoidal holdfast, 10 – 15 cm tall; erect axes feather like (Figs. a, b), 2 – 3 mm thick, pinnately branched up to 3rd order, heavily corticated towards the base, branches alternating in two rows on both sides of main axis, ultimate branches spine-like and polysiphonous at the base and monosiphonous at the distal portion; distal portion dichotomously branched pseudolaterals formed from the middle of the ultimate branches.

Ceramiales, Delesseriaceae

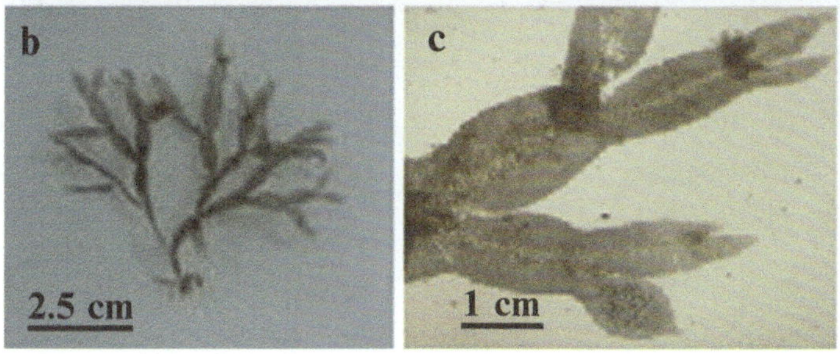

Distribution
Gopnath: N 21° 12.52'; E 72° 06.53'

Caloglossa leprieurii **(Montagne) J. Agardh**

Thallus form and type :	Membranous and constricted
Substratum :	Muddy rocks
Occurrence :	Lower mid littoral zone
Season :	November–April
Abundance :	Moderate
Description :	Plants brownish red to purple red in colour, 7 – 9 cm tall, flat ribbon-like delicate, growing as small tufts, attached by rhizoids arising at the nodes, dichotomously branched with constrictions at each node; internodes with a central midrib and wings on both sides (Fig. c), lanceolate or linear apices often forked (Figs. a, b); tetrasporangia formed at the tips of the thallus by the cells of the wings.
Remark :	It is used in medicine in China particularly as anthelmintic. It is also known as "Ascaris vegetables"

Ceramiales, Delesseriaceae

Hypoglossum heterocystideum **(J. Agardh) J. Agardh**

Thallus form and type	:	Frondose and ribbon type
Substratum	:	Intertidal rocks
Occurrence	:	Infra littoral fringe and Sub littoral zone
Season	:	November–April
Abundance	:	Scanty
Description	:	Plants pale to medium red in colour, 10 – 12 cm tall, delicate, flat, bushy, membranous, densely branched and attached by discoid holdfasts; fronds linear to lanceolate, undivided, margins entire sometimes crisped, about 4 mm in width at its broader part with acute apex; secondary branches arising from the midrib, sometimes abundant proliferations forming from the midrib (Fig. a); marginal cells squarish or rectangular (Fig. b); tetrasporangia in elongated narrow belts on both the sides of the midrib; spermatangia as elongate sori on both the side of the midrib (Fig. c).

Ceramiales, Delesseriaceae

Distribution
Porbandar: N 21° 37.92'; E 69° 36.48'

***Membranoptera* sp**

Thallus form and type	:	Frondose and ribbon type
Substratum	:	Intertidal rocks
Occurrence	:	Mid littoral zone, rocks bordering tide pool
Season	:	November–April
Abundance	:	Scanty
Description	:	Plants light red in colour, flat, 2 – 3 mm broad with poorly developed mid rib, irregularly or pinnately branched (Figs. a, b).

Ceramiales, Sarcomeniaceae

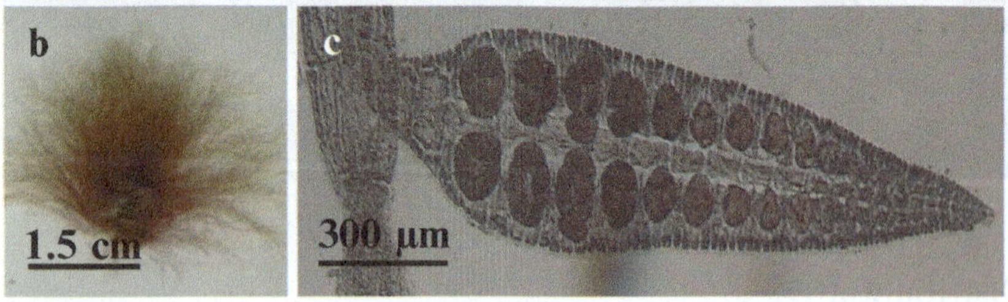

Distribution
Adri: N 20° 57.58'; E 70° 16.76'
Okha: N 22° 28.53'; E 69° 04.41'
Porbandar: N 21° 37.92'; E 69° 36.48'
Samiani Island: N 22° 29.30'; E 69° 05.20'
Veraval: N 20° 54.87'; E 70° 20.83'

Platysiphonia delicata **(Clemente) Cremades**

Thallus form and type	:	Filamentous and polysiphonous
Substratum	:	Intertidal rocks, pebbles, sand
Occurrence	:	Infra littoral fringe, rock pools with coarse sand, stones bordering pools, epiphyte on larger algae
Season	:	November–April
Abundance	:	Moderate
Description	:	Plants dark red in colour, 1 – 4 cm tall, bushy with prostrate and erect filaments, prostrate axes attached to substratum by rhizoids, erect filaments uncorticated giving rise to lateral branches endogenously from the axial cells (Figs. a, b); four pericentral cells formed from axial cells, tetrasporangia produced in two longitudinal rows on wing-like stichidia arising from the lateral branches (Fig. c).

Ceramiales, Rhodomelaceae

Distribution
Okha: N 22° 28.49'; E 69° 04.30'
Porbandar: N 21° 38.24'; E 69° 35.81'
Shivrajpur: N 22° 19.87'; E 68° 56.95'
Veraval: N 20° 54.87'; E 70° 20.83'

Acanthophora dendroides **Harvey**

Thallus form and type	:	Remiform and bushy
Substratum	:	Intertidal rocks
Occurrence	:	Lower mid littoral zone, rock pools with coarse sand
Season	:	November–April
Abundance	:	Moderate
Description	:	Plants dark red to pinkish red in colour, 10 – 15 cm tall, erect with main axis and lateral branches resembling the main axis; branches pyramidally arranged (Figs. a, b); spines absent on main axis, present on branches and ramuli; ramuli attenuated below with occasional spines, spines divergent, conical subulate.

Ceramiales, Rhodomelaceae

Distribution
Okha: N 22° 28.48'; E 69° 04.37'
Porbandar: N 21° 38.24'; E 69° 35.81'
Shivrajpur: N 22° 19.87'; E 68° 56.95'
Veraval: N 20° 54.87'; E 70° 20.83'

Acanthophora nayadiformis **(Delile) Papenfuss**

Thallus form and type	:	Remiform and bushy
Substratum	:	Intertidal rocks, coralline stones
Occurrence	:	Lower mid littoral zone, rock pools
Season	:	November–April
Abundance	:	Moderate
Description	:	Plants dark red to pinkish red in colour, 15 cm tall, attached with the substratum by rhizoids or a basal disc, axes 1 – 2 mm thick, erect, bushy, profusely and irregularly branched; main axes and principal branches covered sparingly with spines, spines pointing upwards; short branches or ramuli with solitary spines (Figs. a, b).

Ceramiales, Rhodomelaceae

Distribution
Okha: N 22° 28.39'; E 69° 03.56'
Porbandar: N 21° 38.24'; E 69° 35.81'
Shivrajpur: N 22° 19.87'; E 68° 56.95'
Veraval: N 20° 54.87'; E 70° 20.83'

Acanthophora specifera **(Vahl) Børgesen**

Thallus form and type	:	Remiform and bushy
Substratum	:	Intertidal rocks, coralline stones
Occurrence	:	Mid littoral zone, rock pools
Season	:	November–April
Abundance	:	Moderate
Description	:	Plants dark red to purple red in colour, up to 20 cm tall, bushy, erect, cylindrical and attached to substratum by irregularly lobed discs (Figs. a, b); main axes without spines, branches irregular or alternate, scarce, branchlets spirally disposed, ultimate short branchlets covered with short spines.

Ceramiales, Rhodomelaceae

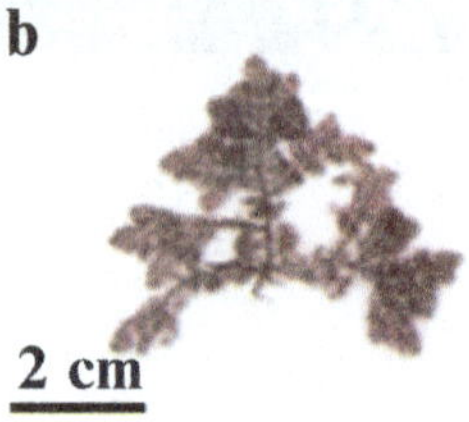

Distribution
Nava Bander: N 22° 45.33'; E 71° 65.23'
Samiani Island: N 22° 29.30'; E 69° 5.20'

Bostrychia tenella **(Lamouroux) J. Agardh**

Thallus form and type	:	Creeping and polysiphonous
Substratum	:	Intertidal rocks
Occurrence	:	Supra littoral zone
Season	:	November–April
Abundance	:	Moderate
Description	:	Plants violet red in colour, up to 3 cm tall with prostrate decumbent axes, attached by discoid heptera developed at the ends of uniseriate rhizoids; rhizoids formed form the pericentral cells; branching pinnate up to 3rd order (Figs. a, b); axes, branches and branchlets corticated except at the terminal parts of the branchlets, which remain monosiphonous.

Ceramiales, Rhodomelaceae

Distribution
Dwarka: N 22° 14.61'; E 68° 57.28'
Okha: N 22° 28.48'; E 69° 04.54'
Porbandar: N 21° 37.92'; E 69° 36.48'
Veraval: N 20° 54.87'; E 70° 20.83'

***Chondria armata* (Kützing) Okamura**

Thallus form and type	:	Remiform and feathery
Substratum	:	Intertidal rocks
Occurrence	:	Lower mid littoral zone and infra littoral fringe
Season	:	November–April
Abundance	:	Scanty
Description	:	Plants violet red to pinkish red in colour, 4 – 10 cm tall, with rhizomatous prostrate portion, attached by conical holdfasts and clusters of rhizoids; erect fronds simple or branched, cylindrical, pinnately branched, ramuli sub-distichously arranged (Figs. a, b), cylindrical or complanate and divided at the distal portions into 2 – 6 ramuli of determinate growth; cystocarps on ramuli, sub-terminal or arising from about the median part.
Use	:	This alga is used in food and medicine (Anthelmintic)

Ceramiales, Rhodomelaceae

Distribution
Nava Bander:
N 22° 45.33'; E 71° 65.23'

Chondria cornuta **Børgesen**

Thallus form and type	:	Remiform and bushy
Substratum	:	Intertidal rocks, coarse sand beds
Occurrence	:	Mid littoral zone
Season	:	November–April
Abundance	:	Moderate
Description	:	Plants dark red in colour, up to 15 cm tall, bushy, main axis slightly branched below and well branched above with long branches at the lower parts and shorter ones at the upper parts (Fig. a); branches cruciform, tapering near the base and at the apex, cystocarps developing on the upper ends of the branches and branchlets, apices of the branches occurring as spines below the cystocarps (Fig. b); tetrasporangia in the upper ends of the branches, branches twisted during growth (Fig. c).

Ceramiales, Rhodomelaceae

Distribution
Dwarka: N 22° 14.61'; E 68° 57.28'
Okha: N 22° 28.46'; E 69° 04.07'
Porbandar: N 21° 37.92'; E 69° 36.48'
Shivrajpur: N 22° 19.87'; E 68° 56.95'
Veraval: N 20° 54.87'; E 70° 20.83'

***Chondria dasyphylla* (Woodward) C. Agardh**

Thallus form and type	:	Remiform and bushy
Substratum	:	Intertidal rocks, coralline stones
Occurrence	:	Mid littoral zone, rock pools, streams
Season	:	November–April
Abundance	:	Scanty
Description	:	Plants dark red in colour, 10 – 20 cm tall, cylindrical, bushy, pyramid like, regularly branched (Figs. a, b); branches single or in pairs or fascicled (whorled), patent, sometimes sub-horizontally extended, clavate with constricted base, cylindrical with truncate apices or obovate with depressed apex; cystocarps sub-apical in short ramuli, single or sometimes more than one on a ramulus, globose to ovate.

Ceramiales, Rhodomelaceae

Distribution
Dwarka: N 22° 14.61'; E 68° 57.28'
Okha: N 22° 28.43'; E 69° 04.09'
Veraval: N 20° 54.87'; E 70° 20.83'

***Digenea simplex* (Wulfen) C. Agardh**

Thallus form and type	:	Remiform and bushy
Substratum	:	Intertidal rocks
Occurrence	:	Lower mid littoral zone and infra littoral fringe, rock pools
Season	:	November–April
Abundance	:	Scanty
Description	:	Plants dark red to brownish red in colour, attached by a basal crust (Figs. a, b), 5 – 15 cm tall, bushy, caespitose, pinnately branched above with alternate branching; branches tapering to acute apices or further ramified, determinate branches corticated at the base; segments with 6 – 8 pericentral cells.
Use	:	This alga is used in medicine as an anthelmintic. Reported to yield bioactive substances

Ceramiales, Rhodomelaceae

Distribution
Adri: N 20° 57.58'; E 70° 16.76'
Okha: N 22° 28.40'; E 69° 04.16'

Herposiphonia secunda **(C. Agardh) Ambronn**

Thallus form and type	:	Filamentous tufts
Substratum	:	Intertidal rocks
Occurrence	:	Supra littoral zone
Season	:	July–November
Abundance	:	Moderate
Description	:	Plants dark red in colour, small, creeping, polysiphonous, attached by unicellular rhizoids arising from the decumbent filaments; branches of indeterminate growth, rudimentary, arising at irregular intervals or fourth or fifth segments (Figs. a, b), usually one or two determinate branches arising between the two indeterminate branches; determinate branches 0.5 – 1 mm in length with 8 – 10 pericentrals in each segments, trichoblast present in the upper portion of the determinate branches.

Ceramiales, Rhodomelaceae

Distribution
Porbandar: N 21° 38.24'; E 69° 35.81'

Laurencia claviformis **Børgesen**

Thallus form and type	:	Filamentous and branched
Substratum	:	Intertidal rocks, coralline substrata
Occurrence	:	Lower mid littoral zone and infra littoral fringe
Season	:	November–February
Abundance	:	Scanty
Description	:	Plants greenish red in colour, up to 1.5 – 3 cm tall, erect; erect branches cylindrical, 1 mm thick in lower parts and about 2 mm thick in upper region; simple or branched (Figs. a, b); branches clavate, irregularly arranged.

Ceramiales, Rhodomelaceae

Distribution
Okha: N 22° 28.45'; E 69° 03.59'
Porbandar: N 21° 38.24'; E 69° 35.81'
Veraval: N 20° 54.87'; E 70° 20.83'

***Laurencia cruciata* Harvey**

Thallus form and type	:	Bushy and cartilaginous
Substratum	:	Intertidal rocks, coralline stones
Occurrence	:	Mid littoral zone and infra littoral fringe
Season	:	November–April
Abundance	:	Dense
Description	:	Plants dark red in colour, 4 – 5 cm high, bushy, uniformly cylindrical from base to apex, pinnately branched, branches distichous or opposite or verticillate (Figs. a, b); ramuli sparse, sterile ramuli cylindrical, truncate, cruciately arranged, fertile ramuli incrassate, swollen, obtuse with unequally developed verrucae.
Remark	:	Some species of *Laurencia* are used in food preparations. They also have known for bioactive substances

Ceramiales, Rhodomelaceae

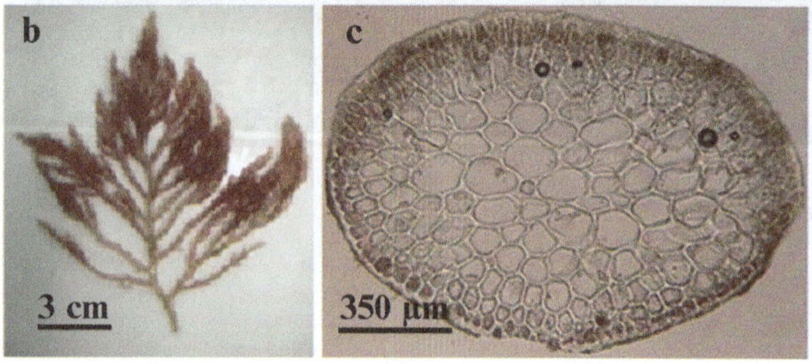

Distribution
Dwarka: N 22° 14.61'; E 68° 57.28'
Okha: N 22° 28.43'; E 69° 03.58'
Porbandar: N 21° 37.92'; E 69° 36.48'
Veraval: N 20° 54.87'; E 70° 20.83'

Laurencia glandulifera **(Kützing) Kützing**

Thallus form and type	:	Bushy and cartilaginous
Substratum	:	Intertidal rocks, stones covered with coarse sand
Occurrence	:	Mid littoral zone, rock pools, rock crevices
Season	:	November–April
Abundance	:	Moderate
Description	:	Plants dark red in colour, 5 – 20 cm tall, terete, cartilaginous, bushy branched alternately with long branches below and short branches towards the apex giving pyramidal appearance (Figs. a, b); ultimate branches sparse very short in the upper part of the main axes and branches; surface cells of the thallus radially elongated or shorter, thick walled cells in the medullary region without lenticular thickenings (Fig. c).

Ceramiales, Rhodomelaceae

Distribution
Okha: N 22° 28.42'; E 69° 03.58'
Porbandar: N 21° 37.92'; E 69° 36.48'
Veraval: N 20° 54.87'; E 70° 20.83'

Laurencia majuscula **(Harvey) Lucas**

Thallus form and type	:	Bushy and cartilaginous
Substratum	:	Intertidal rocks
Occurrence	:	Mid littoral zone, shallow tide pools
Season	:	November–April
Abundance	:	Scanty
Description	:	Plants reddish purple in colour, up to 20 cm tall with 1 – 5 percurrent axes arising from discoid holdfast; densely branched, bushy, pyramidal in appearance (Figs. a, b); branching up to many orders, radial or irregular, branches terete, alternate, ultimate branches many, clavate, constricted at the base.

Ceramiales, Rhodomelaceae

Distribution
Okha: N 22° 28.38'; E 69° 03.56'
Porbandar: N 21° 38.24'; E 69° 35.81'
Veraval: N 20° 54.87'; E 70° 20.83'

Laurencia obtusa **(Hudson) Lamouroux**

Thallus form and type	:	Bushy and cartilaginous
Substratum	:	Intertidal rocks, calcareous stones
Occurrence	:	Mid littoral zone and infra littoral fringe, tide pools
Season	:	November–April
Abundance	:	Dense
Description	:	Plants greenish with pink tips, up to 25 cm height, bushy, caespitose, terete to slightly compressed; branching pinnate, pyramidal, opposite, decussate or often verticillate (Figs. a, b); ramuli cylindrical to sub-clavate slightly shorter and constricted at the base with obtuse truncate apices; superficial cells not radially elongated.

Ceramiales, Rhodomelaceae

Distribution
Okha: N 22° 28.41'; E 69° 03.57'
Porbandar: N 21° 38.24'; E 69° 35.81'
Veraval: N 20° 54.87'; E 70° 20.83'

Laurencia papillosa **(C. Agardh) Greville**

Thallus form and type	:	Bushy, cartilaginous and tuberculate
Substratum	:	Intertidal rocks, calcareous stones
Occurrence	:	Mid littoral zone, shallow tide pools, streams
Season	:	November–April
Abundance	:	Dense
Description	:	Plants dark red to brownish red in colour, 8 – 10 cm tall, erect, attached by discoid holdfasts; many erect axes arise from holdfast; erect axes irregularly branched, cartilaginous, branches covered densely with short wart like tuberculate branchlets (Figs. a, b); superficial cells elongated, palisade-like.

Ceramiales, Rhodomelaceae

Distribution
Adri: N 20° 57.58'; E 70° 16.76'

Laurencia pedicularioides **Børgesen**

Thallus form and type	:	Ramiform tufts
Substratum	:	Intertidal rocks, calcareous stones
Occurrence	:	Mid littoral zone, shallow tide pools
Season	:	July–November
Abundance	:	Scanty
Description	:	Plants rose red in colour, 8 – 15 cm tall, attached by discs; fronds compressed 3 – 4 mm in breadth and 1 mm thick (Fig. a); branching distichous, pinnate, secondary branches irregularly produced, ultimate branchlets 3 mm long, narrow below, broadened at the apical portion, simple or divided into 2 – 3 lobes (Fig. b).

Ceramiales, Rhodomelaceae

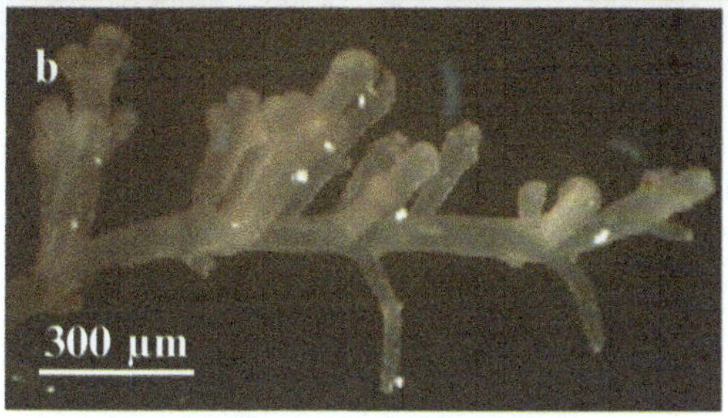

Distribution
Okha: N 22° 28.43'; E 69° 03.58'
Porbandar: N 21° 37.92'; E 69° 36.48'

Laurencia perforata **(Bory) Montagne**

Thallus form and type	:	Bushy, irregularly and intricately branched
Substratum	:	Intertidal rocks
Occurrence	:	Mid littoral zone, exposed rocks
Season	:	November–February
Abundance	:	Scanty
Description	:	Plants purple to green in colour, 2 – 3 cm tall, erect; primary branches curved with attenuated tips which often attached to substratum by rhizoids; secondary branches second, flavate (Figs. a, b); surface cells in the thallus elongated; medullary cells without lenticular thickening.

Ceramiales, Rhodomelaceae

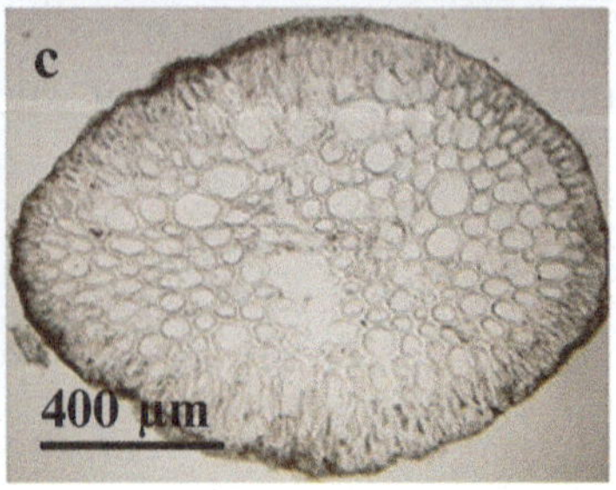

Distribution
Okha: N 22° 28.42'; E 69° 03.58'
Porbandar: N 21° 38.24'; E 69° 35.81'
Veraval: N 20° 54.87'; E 70° 20.83'

Laurencia platyclada **Børgesen**

Thallus form and type	:	Bushy, cartilaginous and compressed
Substratum	:	Intertidal rocks, calcareous stones
Occurrence	:	Mid littoral zone, exposed rocks
Season	:	November–April
Abundance	:	Moderate
Description	:	Plants dark red to pinkish red in colour, up to 10 cm tall, bushy, compressed 1.5 – 2.5 mm in breadth, irregularly pinnately branched; branching marginal, very rarely on the flat surface, alternate, sub-opposite and second with broadly rounded apices (Figs. a, b); secondary branches similar; superficial cells slightly radially elongated, medullary cells without lenticular thickening (Fig. c).

Ceramiales, Rhodomelaceae

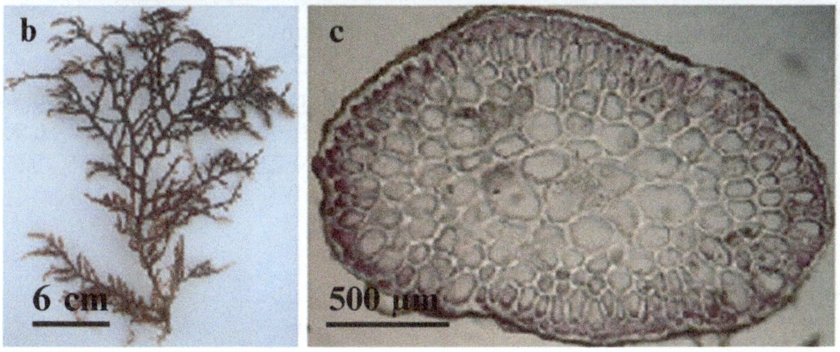

Distribution
Adri: N 20° 57.58'; E 70° 16.76'
Okha: N 22° 28.43'; E 69° 03.58'

***Laurencia* sp**

Thallus form and type	:	Intricate cushions
Substratum	:	Intertidal rocks, calcareous stones
Occurrence	:	Mid littoral zone, tide pools
Season	:	July–November
Abundance	:	Scanty
Description	:	Plants pinkish to dark red in colour, 2.5 cm in height, growing as cushion or patches; fronds pulvinate, intricately, alternately branched; branches cylindrical, tapering from base to apices, covered with small spines (Figs. a, b); superficial cells slightly radially elongated, medullary cells thick walled and without lenticular thickening (Fig. c).

Ceramiales, Rhodomelaceae

Distribution
Okha: N 22° 28.44'; E 69° 03.59'
Porbandar: N 21° 37.92'; E 69° 36.48'
Shivrajpur: N 22° 19.87'; E 68° 56.95'
Vervala: N 22° 16.87'; E 69° 56.75'

Leveillea jungermannioides **(Haring & G. Martens) Harvey**

Thallus form and type	:	Creeping and dorsiventral
Substratum	:	Intertidal rocks, calcareous stones
Occurrence	:	Lower mid littoral zone, tide pools, epiphytic
Season	:	November–April
Abundance	:	Scanty
Description	:	Plants dark red to brownish red in colour, up to 5 cm tall, creeping with dorsiventral axes (Figs. a, b), attached to the substratum by multi-cellular rhizoids arising from the ventral surface, two series of leaf like branches producing alternately along the dorsal surface of the polysiphonous axis (Fig. c); sometimes branches arise from the basal segments of determinate branches; apices of the main axis incurved.

Ceramiales, Rhodomelaceae

Distribution
Adri: N 20° 57.58'; E 70° 16.76'
Okha: N 22° 28.38'; E 69° 04.53'
Porbandar: N 21° 38.24'; E 69° 35.81'
Veraval: N 20° 54.87'; E 70° 20.83'

Lophocladia lallemandi **(Montagne) Schmitz**

Thallus form and type	:	Remiform and feathery
Substratum	:	Intertidal rocks, coralline stones
Occurrence	:	Mid littoral zone, rock pools
Season	:	November–April
Abundance	:	Scanty
Description	:	Plants rose red in colour, 5 – 10 cm tall, polysiphonous (Figs. a, b), segmented; branches exogenously formed in place of trichoblast, each segment producing a single branch initial even before the formation of the pericentral cells; rhizoids formed from the pericentral cells of older segments, branches developing from some branch initial, many other initials transforming into trichoblasts (Fig. c); trichoblasts repeatedly forked with long slender cells; tetrasporangia in short contorted, stichidia produced in the trichoblasts.

Ceramiales, Rhodomelaceae

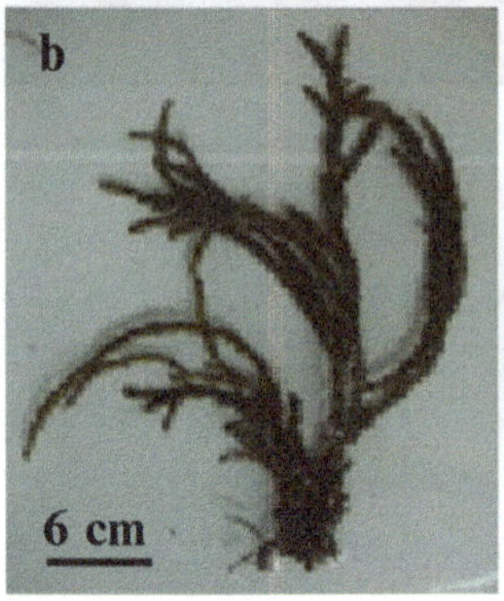

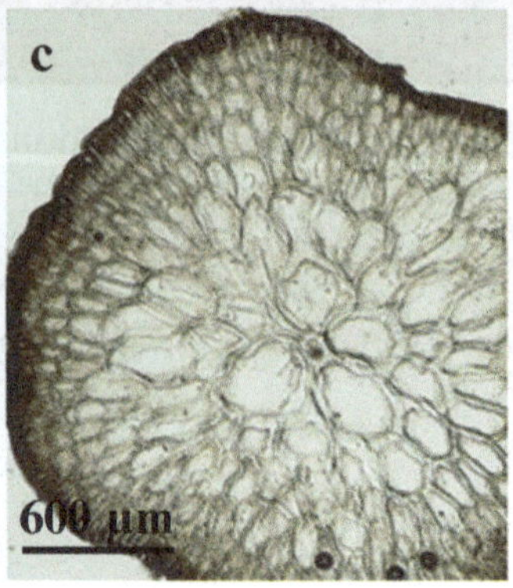

Distribution
Adri: N 20° 57.58'; E 70° 16.76'
Navabander: N 22° 45.33'; E 71° 65.23'

Odontothalia veravalensis **Krishnamurthy et Vijaya**

Thallus form and type	:	Succulent and polysiphonous
Substratum	:	Intertidal rocks, calcareous stones
Occurrence	:	Lower mid littoral zone and infra littoral fringe, exposed rocks
Season	:	November–April
Abundance	:	Dense
Description	:	Plants dark red or brown in colour, 25 – 30 cm tall, erect, cylindrical at the base and becomes progressively compressed above, attached by branched hapteroid discs; axes alternately, profusely branched up to 3rd order (Figs. a, b); main axes and branches corticated, covered with polysiphonous clusters of branchlets with pointed or acute tips; the central axial cells surrounded by four pericentral cells (Fig. c).

Ceramiales, Rhodomelaceae

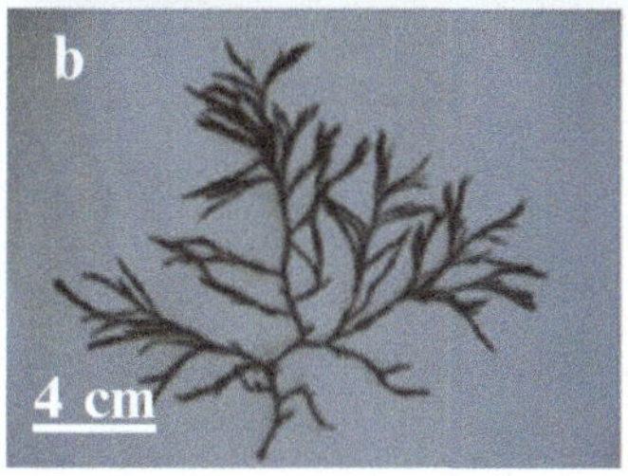

Distribution
Adri: N 20° 57.58'; E 70° 16.76'

***Polysiphonia denudata* (Dillwyn) Greville**

Thallus form and type	:	Remiform and polysiphonous
Substratum	:	Intertidal rocks
Occurrence	:	Mid littoral zone, rocks bordering tide pools, epiphytic
Season	:	November–April
Abundance	:	Scanty
Description	:	Plants dark red in colour, 10 – 15 cm tall, bushy, highly caespitose, expanded, sub-globose, lower filaments thick, dichotomously branched; branches divaricate, upper branches slender with acute axils appearing fastigiated to penicillate (Figs. a, b); segments with 6 – 8 pericentral cells, as long as broad below, longer in the upper portions.

Ceramiales, Rhodomelaceae

Distribution
Okha: N 22° 28.45'; E 69° 04.48'
Veraval: N 20° 54.87'; E 70° 20.83'

Polysiphonia ferulacea **Suhr ex J. Agardh**

Thallus form and type	:	Remiform and polysiphonous
Substratum	:	Intertidal rocks
Occurrence	:	Lower mid littoral zone, shallow tide pools
Season	:	November–April
Abundance	:	Scanty
Description	:	Plants dark red to pinkish red in colour, 4 – 5 cm tall with creeping and erect filaments; creeping filaments attached by ha[i1]pteroid rhizoids (Figs. a, b); branching sparse at the lower parts of the axes, sub-dichotomous and sub-fastigiate in the upper portions with acute angles between the branches, densely corymbose in fertile plants (Fig. c); segments broader than long or as long as broad, gradually tapering towards apex, with four pericentrals, ecorticate, deciduous trichoblasts developing on each segment; cystocarps globose (Fig. d).

Ceramiales, Rhodomelaceae

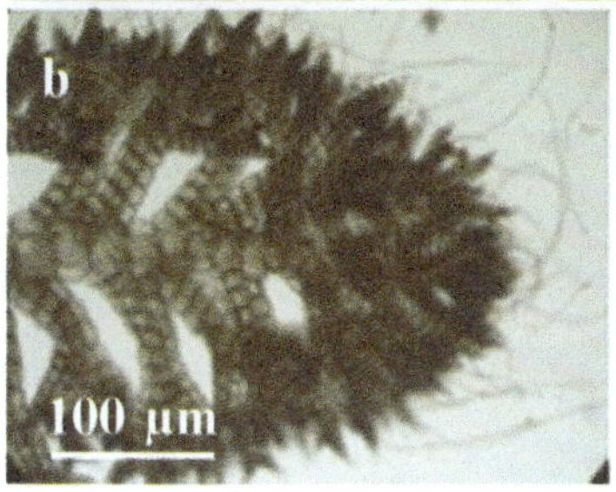

Distribution
Vervala: N 22° 16.87'; E 69° 56.75'

Tolypiocladia glomerulata **(C. Agardh) Schmitz**

Thallus form and type	:	Polysiphonous and spongy tufts
Substratum	:	Intertidal rocks, calcareous stones
Occurrence	:	Upper mid littoral zone, tide pools
Season	:	November–April
Abundance	:	Moderate
Description	:	Plants dark brown in colour, up to 15 cm high, erect and segmented, freely branched and spongy (Fig. a); indeterminate branches at the interval of 2 – 7 or more segments, determinate branches arising from every segment; apical cells elongated with a blunt tip; trichoblast one or two from each forking of the determinate branches (Fig. b).

References

Børgesen, F. 1935. A list of marine algae from Bombay. Kongelige Danske Videnskabernes Selskab, Biologiske Meddelelser 12: 64.

Børgesen, F. 1937a. Contributions to South Indian marine algal flora-I. J. Ind. Bot. Soc. 16: 1–56.

Børgesen, F. 1937b. Contributions to South Indian marine algal flora-II. J. Ind. Bot. Soc. 16: 311–357.

Dixit, S.C. 1968. Species list of Indian marine algae-II. J. Univ. Bombay 36: 9–24.

Iyengar, M.O.P. 1927. Krusadai Island flora. Bulletin of the Madras Government Museum, New Series, Natural History Section 1: 185–188.

Krishnamurthy, V.K. and Johi, H.V. 1970. *A check-list of Indian marine algae*. Central Salt and Marine Chemicals Research Institute, Bhavnagar, Gujarat. p. 36.

Oza, R.M. and Zaidi, S.H. 2001. *A revised check-list of Indian marine algae*. Central Salt and Marine Chemicals Research Institute, Bhavnagar, Gujarat. p. 296.

Sahoo, D., Sahoo, N. and Debasish 2001. *Seaweeds of Indian coast*. A. P.H. Publication, New Delhi. p. 283.

Srinivasan, K.S. 1969. *Phycologia India* (Icones of Indian marine algae). Botanical Survey of India, Calcutta. 1: 52.

Srinivasan, K.S. 1973. *Phycologia India* (Icones of Indian marine algae). Botanical Survey of India, Calcutta. 2: 60.

Untawale, A.G., Dhargalkar, A.K. and Agadi, V.V. 1983. *List of marine algae from India*. National Institute of Oceanography, Goa. p. 42.

Index

The manufacturer's authorised representative in the EU is Springer Nature Customer Service Centre GmbH, Europaplatz 3, 69115 Heidelberg, Germany. If you have any concerns regarding our products, please contact ProductSafety@springernature.com

Printed and bound by CPI Group (UK) Ltd, Croydon, CR0 4YY
30/09/2026
02225516-0002